# SpringerBriefs in Applied Sciences and Technology

## Automotive Engineering : Simulation and Validation Methods

**Series editors**

Anton Fuchs, Graz, Austria
Hermann Steffan, Graz, Austria
Jost Bernasch, Graz, Austria
Daniel Watzenig, Graz, Austria

W0036967

More information about this series at http://www.springer.com/series/11667

Daniel Watzenig · Bernhard Brandstätter
Editors

# Comprehensive Energy Management—Eco Routing & Velocity Profiles

 Springer

*Editors*
Daniel Watzenig
Virtual Vehicle Research Center
Graz
Austria

Bernhard Brandstätter
Virtual Vehicle Research Center
Graz
Austria

ISSN 2191-530X         ISSN 2191-5318  (electronic)
SpringerBriefs in Applied Sciences and Technology
Automotive Engineering : Simulation and Validation Methods
ISBN 978-3-319-53164-9         ISBN 978-3-319-53165-6  (eBook)
DOI 10.1007/978-3-319-53165-6

Library of Congress Control Number: 2017932086

Printed on acid-free paper

This Springer imprint is published by Springer Nature
The registered company is Springer International Publishing AG
The registered company address is: Gewerbestrasse 11, 6330 Cham, Switzerland

# Contents

## Bernhard Brandstätter[1] and Daniel Watzenig[1] on Behalf of the Cluster of 4th Generation Electric Vehicles

This book is organized in two volumes (the contents of both volumes are listed at the end of this introduction):

- Volume 1: "Comprehensive Energy Management—Eco Routing & Velocity Profiles"
- Volume 2: "Comprehensive Energy Management—Safe Adaption, Predictive Control and Thermal Management"

## Comprehensive Energy Management

Energy management plays a central part in today's vehicles, especially for battery electric vehicles, where a limited number of charging possibilities and time-consuming charging processes lead to range anxiety of the users. This can be considered as an important factor (apart from the increased cost of electrical vehicles compared to conventional ones) that prevents larger number of fully electric vehicles on the road.

Thus comprehensively treating energy and controlling it is of uttermost importance.

This book provides findings of recent European projects in FP-7 grouped in a cluster named "Cluster of 4th Generation Electric Vehicles" but also gives insight into results from ongoing H2020 projects related to energy management.

---

[1]Virtual Vehicle Research Center, Inffeldgasse 21a, Graz, Austria

Since fuel cell technologies are gaining more attraction again, the last section of the book gives an overview of the state of the art in this field what PEM[2] fuel cells is concerned.

# Cluster of 4th Generation Electric Vehicles

The Cluster "4th Generation EV" was set up late 2013 by the European projects INCOBAT, iCOMPOSE and eDAS, with the purpose to synchronize and cojointly promote the R&D topics on electric vehicles. By growing to a total of six projects with the FP7 projects Batteries2020, IMPROVE and SafeAdapt, the cluster also enlarges its networks and range of influence on the European electric vehicle community (Fig. 1).

The projects within the cluster are focusing on the following goals:

- Batteries 2020+: improve performance, lifetime and total cost of ownership of batteries for EVs
- eDAS: Holistic energy management for 3rd and 4th generation EVs.
- iCOMPOSE: Integrated Control of Multiple-Motor and Multiple-Storage Fully Electric Vehicles.
- INCOBAT: Innovative Cost Efficient Management System for Next Generation High Voltage Batteries.
- IMPROVE: Integration and Management of Performance and Road Efficiency of Electric Vehicle electronics.
- SAFEADAPT: enrich networked embedded systems in e-vehicles.

Uniting more than 40 partners from 12 countries all over Europe, including 7 OEMs, with an overall budget of more than 36 million Euros, the impact of the cluster on the next generation of electric vehicles keeps on growing.

The "4th Generation EV" cluster is organized around the three following working groups:

- Comprehensive energy management
- Central computing platform
- Potential of electrification

---

[2]Proton exchange membrane

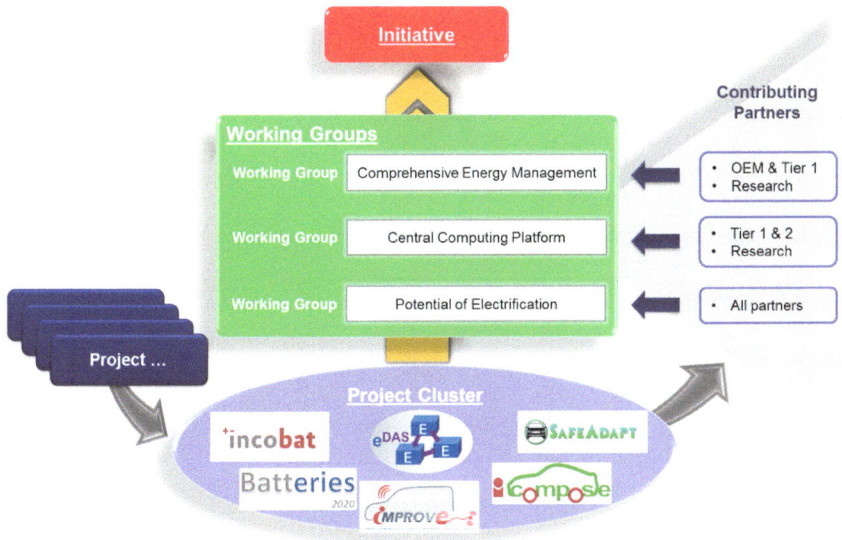

**Fig. 1** Cluster of 4th Generation Electric Vehicles

Some of the cluster projects will end in 2016 but bridging to H2020 projects has already begun. As an example the H2020 project OPTEMUS can be mentioned, where the comprehensive energy aspect is widened by new technologies like heating panels and energy harvesting technologies with strong focus on thermal comfort sensation inside the cabin (which plays a very important role in overall energy consumption).

## Organization of this Book

The chapters of this book are organized under five different groups: ECO driving and ECO routing covers different approaches for optimal speed profiles for a given route (mostly interconnecting with cloud data); model-based functional safety and fault-tolerant E/E architectures; advanced control making use of external information (from a cloud) as well; thermal management as a central part for energy optimization and finally some aspects on fuel cells.

These subject areas with their chapters (chapter titles in italic) are listed below:
**Volume 1**:

- ECO Driving and ECO Routing

  - *Aspects for Velocity Profile Optimization for Fleet Operated Vehicles*: on-board and off-board optimization including cloud communication

- *Semi-Autonomous Driving Based on Optimized Speed Profile*: different controllers including model predictive control
- *Design of Vehicle Speed Profile for Semi-Autonomous Driving: energy consumption optimization for different driving conditions*
- *Energy-Efficient Driving in a Dynamic Environment*: considers energy optimal velocity profiles in the presence of other traffic participants and overtaking possibility
- *Model-Based Eco-Routing Strategy for Electric Vehicles in Large Urban Networks*: energy consumption model that considers accelerations and road infrastructure

**Volume 2**:

- Safety Aspects
  Addressing fault-tolerant approaches of automotive energy-efficient E/E architectures and model-based functional safety engineering in

  - *Safe Adaptation for reliable and Energy-Efficient E/E Architectures*
  - *Model-based functional safety engineering*

- Advanced Control

  - *Model predictive control of highly efficient dual mode energy storage systems including DC/DC converter*
  - *Predictive energy management on multi-core systems*: first approach to solve a reference speed tracking problem on a multi-core platform in real time

- Thermal Management

  - *Holistic thermal management strategies for electric vehicles*: including some rudimentary cabin comfort issues
  - *Heat pump air conditioning systems for optimized energy demand of electric vehicles*

- Fuel Cells

  - *Thermal management of PEM fuel cells in electric vehicles*

The aspects within the field of comprehensive energy management are too numerous that all of them could have been addressed in this book (aerodynamics and adaptive control of aerodynamic features could be mentioned in this context as an example). We think, however, that important key enabling elements for optimal energy management taking the environment and context into account have been collected in this book.

We cordially acknowledge all authors and co-authors for their efforts and looking forward to next steps in future projects.

Graz, Austria                                                                    Daniel Watzenig
December 2016                                                        Bernhard Brandstätter

# Aspects for Velocity Profile Optimization for Fleet Operated Vehicles

Pavel Steinbauer, Jan Macek, Josef Morkus, Petr Denk, Zbyněk Šika and Florent Pasteur

## 1.1 Introduction and Problem Formulation

The presented approach was developed by members of EU FP 7 project IMPROVE consortium. IMPROVE is focused on commercial electric light duty vehicles, which are operated in urban surroundings (either within a fleet or as single-operated vehicle). The project aims at improvement of a daily duty range, using realistic range assessment and following adaptive optimization. Limited range of electric vehicles is a main obstacle for their wide-spread use. The current progress of information and communication technologies (ICT) can extend the achievable range using the best energy management, even if it is still limited by physics of driving resistances [1, 2].

Current electric vehicles are complex mechatronic devices. The pickup vehicles of small sizes are currently used in transport considerably. They often operate within a repeating scheme of a limited variety of tracks and larger fleets. Thanks to mechatronic design of vehicles and their components, and availability of high capacity data connection with computational centers (clouds) and the data storage capacity in them, there are many means how to optimize their performance, both by planning the trip and recalculations during the route. The data capacity may be used in both directions—as a source of vehicle-to-infrastructure communication, V2I, or as storage capacity for vehicle history.

P. Steinbauer (✉) · J. Macek · J. Morkus · P. Denk · Z. Šika
Faculty of Mechanical Engineering, CTU in Prague, Prague, Czech Republic
e-mail: Pavel.Steinbauer@fs.cvut.cz

F. Pasteur
Siemens Industry Software S.A.S. DF PL STS CAE 1D, Lyon, France

© The Author(s) 2017
D. Watzenig and B. Brandstätter (eds.), *Comprehensive Energy Management—Eco Routing & Velocity Profiles*, Automotive Engineering: Simulation and Validation Methods, DOI 10.1007/978-3-319-53165-6_1

Comprehensive energy management for electric vehicles has to take into account extended map-based route conditions (especially legal speed limit, safety speed limit due to local curvature of turns, slope, horizontal direction, basic adhesion factor typical for road surface, location of charging stations, traffic conditions from infrastructure information system, traffic density defined, e.g., by achieved average speed, waiting time for charging at different charging stations, weather conditions, especially temperature, rain or snow, wind speed and its direction) and operation conditions (target locations—more targets are typical, e.g., for commercial delivery vehicles with loads (changing typically along the route).

Optimum trip driving schedule for achieving pre-defined operation conditions [the target(s) with delivery of the load(s)] is usually defined by the lowest energy consumption schedule while maintaining pre-defined constraints (time of reaching targets below the limit, temperature inside a vehicle kept between suitable minimum and maximum, safety factor of using adhesion limit (skid safety factor) below the limit, temperature of a motor and other electric equipment below the limit, battery state-of-charge (SOC) above the limit.

The important feature of optimum trip design is not only its predictive phase but its adaptiveness, i.e., the re-optimization during a trip if circumstances (e.g., traffic density, weather, etc.) change. The on-board diagnostics tools (e.g., for more accurate load determination) may be used as inputs, as well. In this phase, the computational power of the cloud can be used as well, but properly calibrated, on-board simpler simulation tools may be used independent on availability of cloud services with advantage. This is why the model hierarchy should be defined. The simple models may be calibrated in advance using data from previous operation history of the same vehicle or by optimizing the parameters of the simple model by comparison to the results of full-size vehicle model during pre-trip optimization. The simpler model may use simplified optimization based on generalized set of limited number of optimization parameters, as well. The future progress of electric vehicle operation economy can be done by using the battery durability management and on-line battery and vehicle diagnostics, based on evaluation of vehicle history data.

The following description of the problem solver is divided into route description suitable for use with a vehicle model for simulation of driving and heating/air-conditioning (HVAC) energy consumption. Algorithms for full pre-trip optimization will be described together with their on-board implementation, and closed-loop control design will close the current section.

Although many aspects of this opportunity were already addressed [1, 2], this article shows an approach developed to further increase the range of e-vehicle operation. It is based on prior information about the route profile, traffic density, road conditions, past behaviour, mathematical models of the route, vehicle and dynamic optimization. The most important part of the procedure is performed in the cloud, using both computational power and rich information resources. Suitable route discretization into sections is the most important part of the algorithm. The various information resources are used. The accumulated experience coming from fleet operation is also very important. Methods for automation of this procedure are presented. Subsequently, feasible initial values of section parameters are found using heuristic rules devised from good driver's practice and backward calculation based on dynamic programming principles. Designed velocity profile is further optimized based on simplified, but very fast energy consumption models, verified and fine-tuned on detailed simulation model of the vehicle. The velocity profile is updated when requested and finally loaded into on-board control unit. A model based predictive controller (MPC) is used to keep the vehicle with its driver efficiently on defined track. The proposed strategy is verified in a simulation environment and prepared to be implemented on the test vehicle and the cloud system.

## 1.2 IT Architecture and Overall Algorithm

The system consists of several main components. The basic control strategy is defined and optimized before the journey of the vehicle starts using computational power and information resources in the cloud (Fig. 1.1).

The route definition is obtained from navigation algorithm and map data (HERE —an open location platform currently developed and maintained by consortium of mainly European automotive manufactures [3]), complemented with altitude information and tuned by history data about previous experience with given route. The output of this optimization is the optimal velocity route profile (OVRP), which complements current classical navigation data sets. This route description is based on natural discretization of the route into sections and setting velocity parameters for each section. These OVRP data are transferred into EVCU (Electric Vehicle Control Unit) and used as forward controller component. On-board sensor information (mainly forward distance, state of charge, torque, battery temperature etc.)

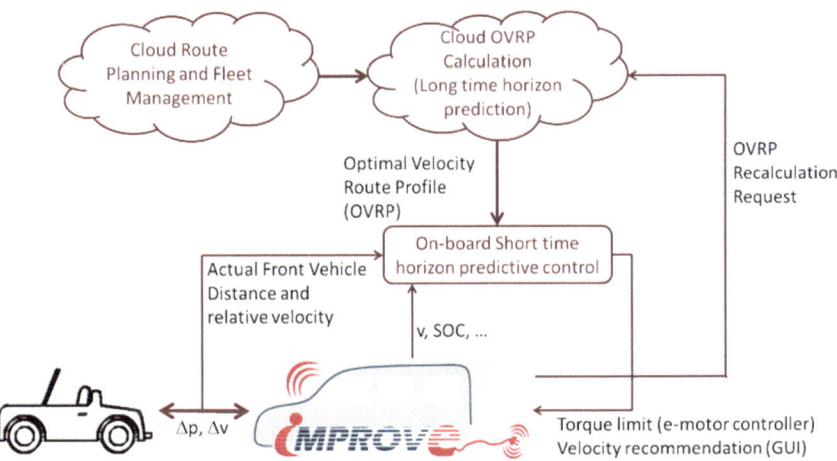

**Fig. 1.1** Software architecture

are used in closed loop control component based on a predictive and adaptive control algorithm (PACA). In case of substantial difference between pre-optimized velocity profile and actual traffic situation, the high level OVRP calculation is repeated in the cloud. The required velocity is displayed as assistance to the driver via LCD display and used for recommendation of torque limit for the torque manager component of the vehicle.

The information about vehicle behavior during the ride is stored and transferred into cloud data storage for future planning. The velocity route planning is thus improved gradually as the system is used.

## 1.3 Route Segmentation Techniques

The route description, which is obtained from route planning algorithm [3], is processed for future calculations into sections. The section is defined as part of the route, which has constant or only slightly changing properties, especially route slope and maximum velocity. The maximum velocity is obtained as minimum from legal limit (which is part of route description) and maximum velocity (physically achievable and acceptable for passengers). The latter is calculated from route curvature, adhesion limit and comfort coefficient based on force equilibrium (Fig. 1.2).

$$\sum Fx = N \times \sin\theta + \mu_s \times N \times \cos\theta$$
$$\sum Fy = 0 = N \times \cos\theta - \mu_s \times N \times \sin\theta - m \times g \qquad (1.1)$$

The maximum velocity in the curve is given by the maximum radial acceleration derived from adhesion limit, which is adjusted by passenger comfort coefficient $k_{comfort}$

$$v_{comfort} = \sqrt{r \times g \times \mu_s \times k_{comfort}} \qquad (1.2)$$

The reason is both security and comfort level of the passengers who are exhibited to lateral forces.

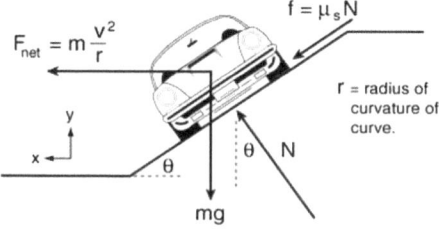

**Fig. 1.2** Forces acting on the vehicle riding through curve

The radius r of each route point is determined by circular regression through several neighboring points (Fig. 1.3b). The number of neighboring points determines filtering level, which covers point coordinates uncertainty. Also, unlike the usual digital filters, such filtering takes into account non-equidistant character of the positional data.

The slope is determined in similar manner (Fig. 1.3c), using regression by straight line through several neighboring points. The proper non-equidistant point distance filtering is ensured again.

The new section boundary is defined as follows:

1. whenever the legal speed changes to another speed limit or
2. whenever the speed is reduced or increased over the legal limits due to road profile or traffic or obstacles. The section speed is calculated for each route section as the average speed over the section.
3. Whenever the slope of the route changes more then defined tolerance.

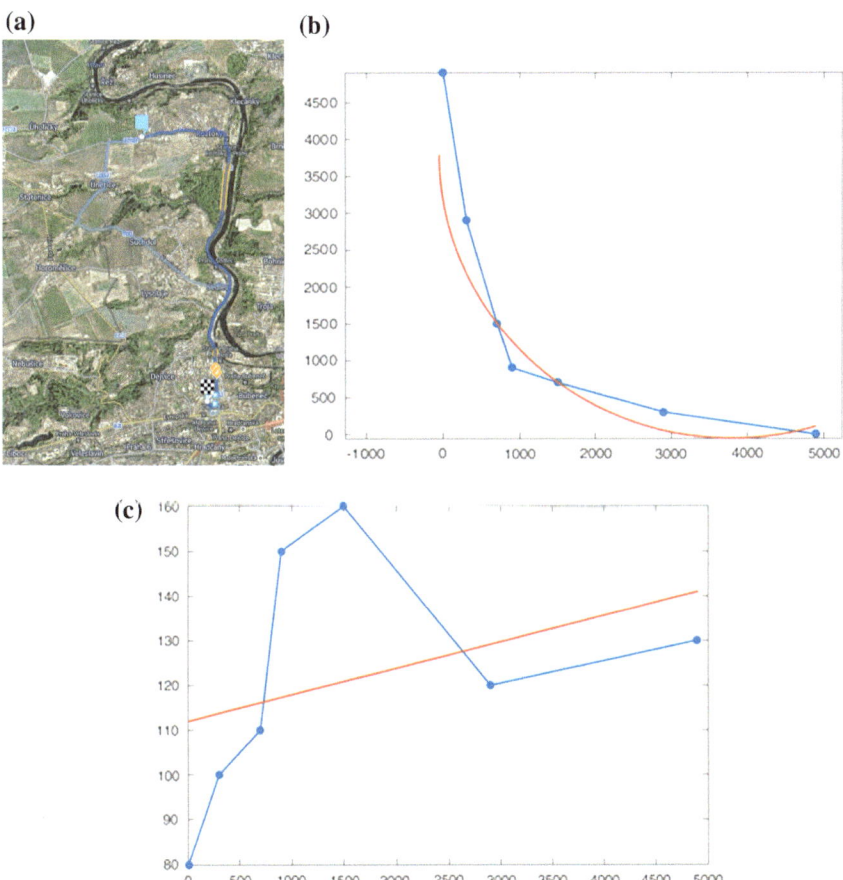

**Fig. 1.3**  Route from navigation (**a**), curvature (**b**) and slope determination (**c**)

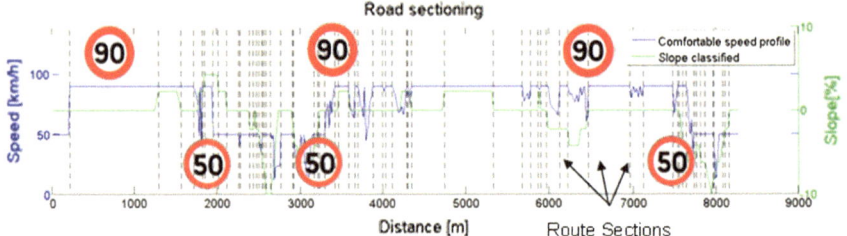

**Fig. 1.4** Route section definition

The process results into sequence of sections with constant main properties (Fig. 1.4).

To learn from accumulated fleet experience, various track records have to be combined. The various data sources blending must be carefully carried out, as there is different number of positional points from different records.

The consistency check is done first to exclude false measurements (it is based on maximum velocity between points). Next step is sorting out points by checking the direction of the vectors between subsequent points (Fig. 1.5a).

For optimization of optimal velocity profile, in each section four phases of vehicle driving mode are defined (Fig. 1.5b): the phase of acceleration, the phase with constant velocity of the vehicle, the phase of coasting and the phase of deceleration. The order of phases cannot be changed, but some of them are skipped in particular cases, i.e. they have zero length. E.g. the coasting and deceleration will not occur if consecutive section has higher constant velocity.

Each section is defined by a set of parameters that describe a given section. This set of parameters should be composed only by mutually independent parameters. Otherwise, the number of parameters can be reduced and parametrization completed with appropriate number of constraints.

The section velocity profile is fully described by set of five parameters $[v_1, a_a, s_a, s_d, a_d]$, their meaning can be seen in the Fig. 1.5b. The terminal velocity of deceleration $v_4$ is not included in the set of parameters of the section as a separate parameter as is $v_1$ of next section. The total trajectory of the vehicle is composed of successively connected sections. Two successive trajectory sections are interconnected in a connection point, where the terminal velocity $v_4^i$ of first section $i$ is also the input velocity $v_1^{i+1}$ in the next section $i + 1$.

The set of parameters creates the velocity profile as follows

$$v_2 = \sqrt{2a_a s_a + v_1^2}, \quad v_3 = \sqrt{2a_d s_d + v_4^2}$$
$$s_{coa} = f(v_2, v_3), \quad s_{con} = s - s_a - s_d - s_{coa}(x, u, t) \tag{1.3}$$

The main advantage of such parametrization is that it allows the zero length of section phases without getting into numerical and mathematical problems.

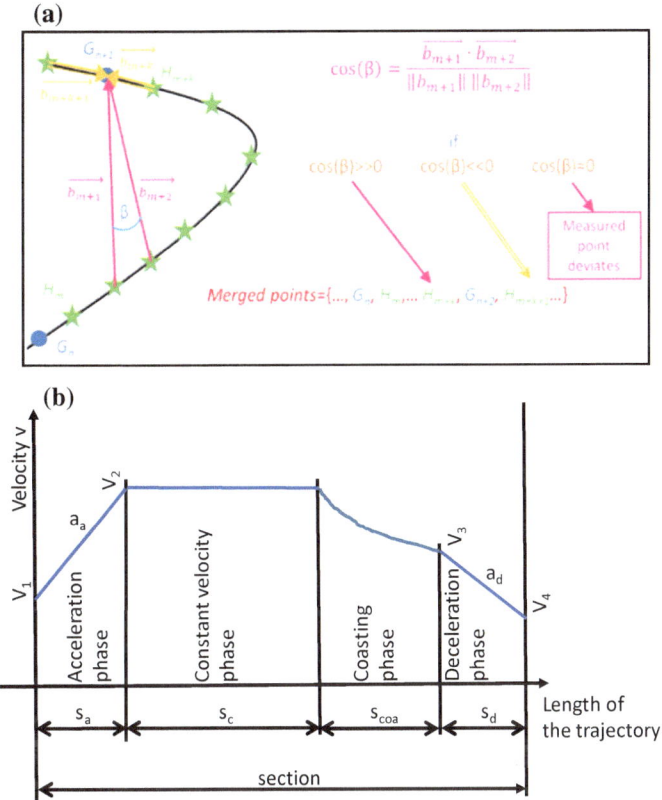

**Fig. 1.5** History data blending strategy (**a**), four-phases section description (**b**)

## 1.4 Model Design and Calibration

Optimal velocity route profile, controller design and results verification are based on mathematical models. Several models of different complexity are used. The LMS software was used to develop a full, complex vehicle model for range estimation. However, such complex model is computationally demanding and cannot be used for repeated calculation in optimization algorithms. So a simplified energy consumption model, mostly based on algebraic relations, was developed for this purpose. Finally, control design was based on a grid of linearized models.

The Vehicle Energy Management (VEM) simulator is a LMS.IMAGINE.Lab Amesim 1D virtual model of the Fiat Doblo electric vehicle (Fig. 1.6a, b). This simulation platform is well suited to carry on global vehicle energy consumption evaluation [2, 4]. It is a range prediction model, containing the main subsystems: vehicle dynamics 1D model with front and rear axles, model of an electric machine and DC converter using real measurement data-files, which define the losses, the minimum and the maximum torques, high voltage battery, modelled as a quasi-static equivalent electric circuit model, defined with the datasheet measure information, taking into account the OCV and resistance dependence as a function of SOC and temperature and simplified aging model. HVAC system is defined as a simplified reduced model, the vehicle auxiliary consumers i.e. all the vehicle electric equipment that belongs to the low voltage on board network of the vehicle (12 V) and finally a vehicle control unit (VCU), which computes the motor torque demand according to the vehicle state, the pedal position, battery SOC, inverter power limitation (high frequency control dynamics such as ESP have been neglected).

The LMS model has been modified to be integrated into the MATLAB/Simulink environment. A coupling block is added to define the quantities to be exchanged between LMS-Amesim and Simulink.

However, the real vehicle as well as its mathematical model are highly non-linear, especially with respect to velocity and state of charge of the battery pack. Therefore, a linear controller resulting from MPC (model predictive control) design procedure at one set point of the model does not work well at remote state space positions. Thus, a set of linear models is created for a grid of set-points.

The new Doblo simulation model has to be specifically modified in order to perform the linear analysis. The vehicle speed and battery SOC (state of charge) are chosen to define the linearization grid of the system (Fig. 1.7). The driver model is removed and replaced by the torque request.

The simulation model is integrated during 0.1 s in order to reach the steady state defined by this set point. In order to check the validity of this approach, each linearized model has been compared to the original non-linear model. Based on this verification, the linearization grid density has been adapted to ensure sufficiently accurate results.

The fast vehicle model used for optimization is based on the analysis of energy flow and losses. The simple algebraic equations must be derived to achieve rapid evaluation. So, it is necessary to include analysis of vehicle components into the calculations and optimization of control strategy. In principle, the most important losses are depicted in Fig. 1.8. The cost function is based on sum of used energies, mainly

- rolling resistance—$E_f$
- slope resistance—$E_s$
- acceleration resistance—$E_a$
- air resistance—$E_v$
- additional unit of vehicle (trailor)—$E_t$
- heating/air conditioning as electric power input to a heat pump—$E_h$

All differential equations of motion or of thermal energy conservation are integrated with constant parameters for a single route section. In the case of HVAC systems, the most general case with heat pumps for heating or cooling is implemented together with a simple heat balance model of a car body (driver's room

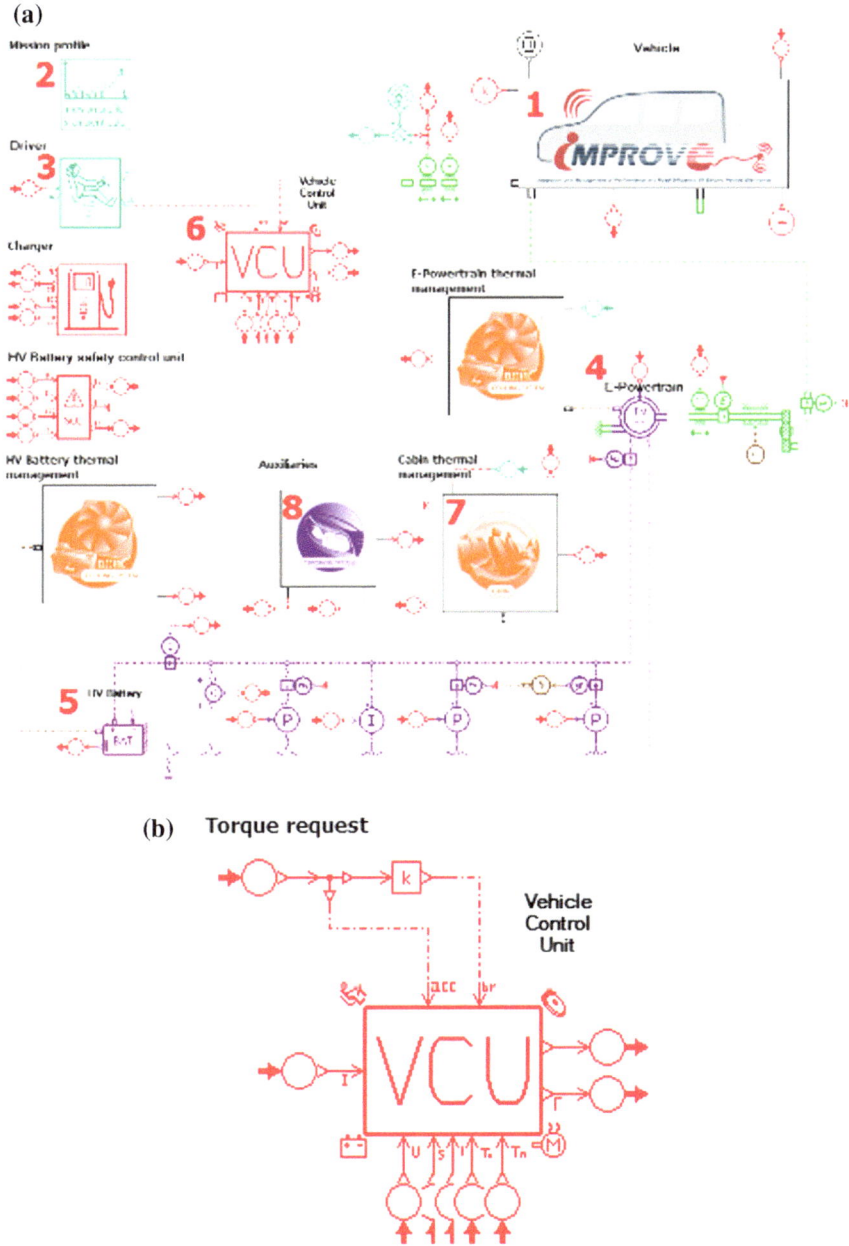

**Fig. 1.6** LMS-amesim vehicle energy management model (**a**), SIMULINK coupling (**b**)

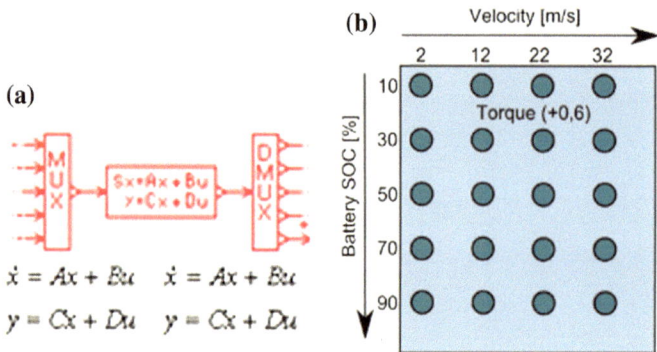

**Fig. 1.7** Linear model form (**a**), linearization in the grid (**b**)

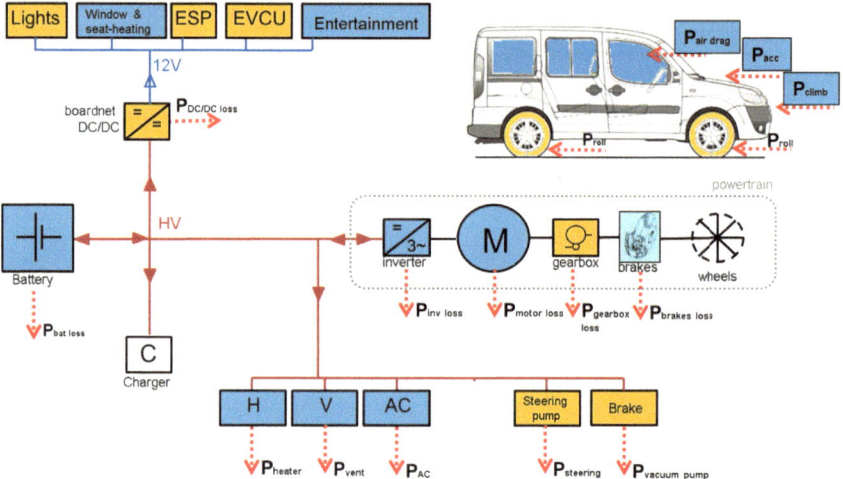

**Fig. 1.8** Vehicle architecture and external forces acting to a vehicle for energy management optimization

temperature) and a windshield (defrosting or de-fogging). The heat pumps are simulated by Carnot efficiency corrected by a constant multiplier to real heat or cooling factor. The possibility of extracting thermal energy from the cooling circuit for electric parts of powertrain is envisaged. In this way, both constraints of electric components maximum temperature and as suitable driver's compartment temperature may be taken into account.

Payload consisting of cargo and crew may vary during the trip. It is thus reasonable to on-line monitor vehicle payload state. Necessary additional sensors should be inexpensive, easy to install, or even completely avoided.

Two methods were developed. The first one is based on well-known equations of motion for longitudinal dynamics for acceleration with constant internal motor torque $M_M$ and deceleration for coast-down.

Unknown rolling resistance and rather uncertain measurement of local slope of a road can be excluded. Then for unknown payload mass yields

$$m_L = \frac{\frac{i_g}{r_w}\left(M_M - M_F + M_F'\right) + \frac{c_d S_v \rho}{2}\left(w'^2 - w^2\right)}{a_d + a_a} - m_v(1 + \delta) \qquad (1.4)$$

The $m_v$ and $m_L$ are masses of the vehicle and the payload respectively, $c_r$ is coefficient of rolling resistance, $c_d$ is drag coefficient, $\alpha$ is road slope, $\delta$ is reduction of rotating masses to empty vehicle mass, $a_a$ is acceleration, $a_d$ is deceleration, w is mean velocity during acceleration phase, $w'$ is mean velocity during coast-down, $M_F$ is friction torque of electric motor+ transmission if motor active, resp. $M_F$, if motor is inactive during coast down, $i_g$ stands for transmission ratio.

Another approach is based on the fact that change of sprung mass will change eigen frequency of vertical mode. The eigen frequency can be determined from analyses of vertical vibration motion measurement. This can be accomplished by suitable acceleration transducer, DAQ hardware and computer post-processing. As the computational power of current signal processors is quite high, even complex calculations, including Fast Fouriér Transform algorithms (FFT), can be done easily.

The eigen frequency of vertical motion mode (typically up to 4 Hz) changes, however, moderately due to high stiffness of springs and tires, not more than 0.3 Hz with mass change. So the measured signal contains non-negligible frequency components in the range of 0–1000 Hz. Signal conditioning must be adjusted to it. Also, the level of acceleration of sprung mass (chassis) is usually up to 2 g. Quite high resolution of frequency determination $\Delta f$ is necessary, based on performed analysis at least 0.01 Hz. However, sufficiently long time window $T = 1/\Delta f$ can be easily measured during the ride.

The rolling resistance is often found from coast-down deceleration, which can be measured similarly to active motor acceleration. The typical values for the current commercial vehicles are between 0.3 and 0.6 m s$^{-2}$ (according to the load and vehicle speed—drag). Unlike vehicle mass, the procedure is not robust, as follows from Eq. (1.7) written here for a horizontal plane movement

$$c_R = \frac{-\frac{i_g}{r_w}M_{F1}' + [m_v(1 + \delta) + m_L]a_{d1} - c_d S_v \rho \frac{w_1'^2}{2}}{g(m_v + m_L)} \qquad (1.5)$$

The vehicle mass, with all uncertainties described above, and the drag are inputs into the equation. The friction torque may be added to the rolling resistance coefficient due to small dependence on driveline speed.

The large issue is caused by generally unknown slope. In laboratory tests it is avoided using two opposite directions of coast-down at the same vertex of a road, which is not suitable for daily diagnostics.

The possibility is to check the horizontal plane by leaving a vehicle with released brakes for a few seconds staying on a test plane and checking whether some movement occurs. If it is not the case, procedure of both active acceleration test and coast-down deceleration measurement may be started. The active acceleration may be used with known motor torque for checking the coast-down results, but this check cannot avoid the slope influence (both accelerations will be shifted by the same slope acceleration component g.sinα). The longitudinal axis angle measurement would be of advantage if calibrated by horizontal plane position.

## 1.5   Algorithms for Full Pre-trip Optimization

The optimal velocity route profile is developed by a 3-phase algorithm. The first phase determines a feasible, fastest velocity profile, based on available motor power and braking capabilities of the vehicle. Then velocity profile, based on heuristic rules which follow from good driver practice, is calculated. Finally, dynamic optimization takes place and fine tuning of velocity profile based on fast vehicle model.

Backward-Forward Heuristic Calculation

The search for maximum velocities starts from the end of the route. It enables to determine maximum velocities, at which the vehicle can slowdown into following section velocity limit.

The procedure with examples of possible series of sections is demonstrated in Fig. 1.9. The predictor of maximum outlet velocity from the current section guarantees the successful braking to inlet velocity of all further sections.

The corrector finds real $w_{out,i} \leq w_{out,max,I}$ taking velocity limit and possible positive acceleration, using opposite direction (Fig. 1.10). If optimum velocity is greater than inlet and outlet velocities (it must not be less than those because of predictor step), acceleration to the limit is considered.

The same backward approach can be used for optimizing charging points during a long trip.

Heuristic optimization can be done with very limited number of independent variables, based on generally valid dimensionless multipliers (e.g., reduction ratio for section speed—except for very low speed limit, which is used fully, reduction ratio for maximum motor power, maximum acceleration or deceleration reduced against adhesion limit, normalized power available for acceleration, etc.). The time constraint may be fulfilled explicitly using more complicated logical block for a speed governor. These features accelerate optimization procedure, which is important especially for adaptive re-optimization.

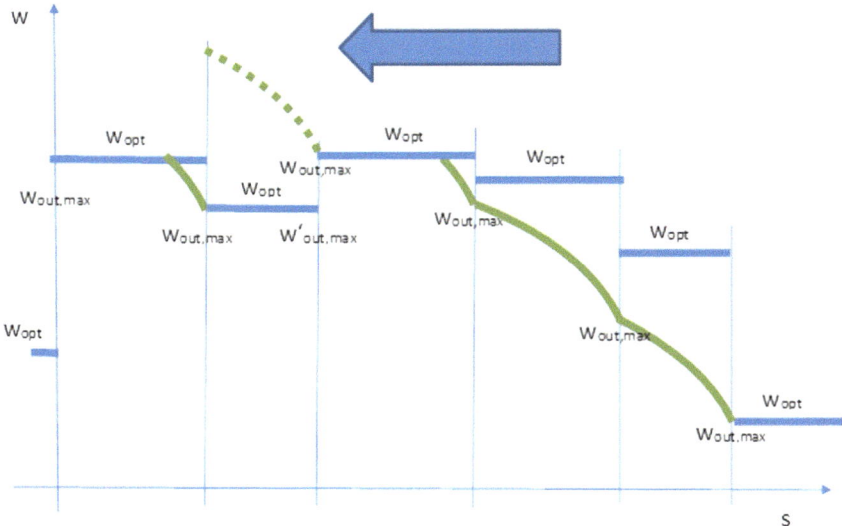

**Fig. 1.9** Backward step: search to find maximum velocity which enable to slow down to limit of following section

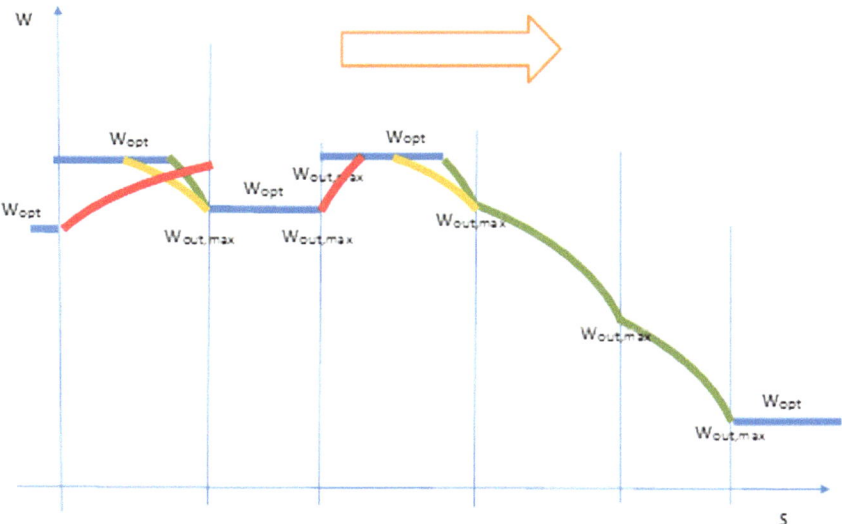

**Fig. 1.10** Forward step: maximum achievable velocity based on max. torque and acceleration

Dynamic optimization

Dynamic process optimization is based on path and vehicle model discretization. It is a full pre-trip optimization. All parameters of velocity profile are tuned respecting the constraints (e.g. maximum velocity in the section or continuity between subsequent sections and mainly maximum time of travel). Starting set is provided by previous heuristic (backward-forward) algorithm.

The following approach is used: The initial velocity profile is obtained by using a-priory knowledge about the system. It is optimized by dynamic system optimization methods, which were developed for optimization of complex and non-linear dynamic processes. They are based on discretization of the state trajectory and replacement of highly non-linear dynamic model by algebraic functions.

The optimization criterion (cost function) is the overall consumed energy for whole route. It is calculated by simplified energy consumption model (mainly algebraic) for each section.

The objective function is a function of constant parameters of vehicle and road and optimization parameters, whose number is the product of the number of sections along the selected trajectory and the number of optimization parameters used to describe the individual phases of the section. For such defined optimization problem and the objective function, the interval of values of the objective function is limited by local and global validity conditions, named "optimization conditions". The local optimization conditions are defined for individual sections of the selected trajectory. They limit the mutual relation of optimization parameters in each section and prescribed fixed length of each section. Global optimization conditions define the maximum and minimum values of optimization parameters, define the continuity of the velocity profile at the endpoint of all sections of the trajectory and the total time of travel of the vehicle along the selected trajectory. All of these conditions form constraint parameter space to be searched.

Dynamic optimization of the given problem is based on optimization technique called "Trust-Region Methods for Nonlinear Minimization" [5]. This method is based on the principle of replacing the objective function f in a neighborhood of an initial estimate by approximation functions. For this approximation function must hold the local and global optimization conditions. Optimization problem approximating function and optimization conditions are defined:

$$\min\left\{\frac{1}{2}s^T H s + s^T g, \|Ds\| \le \Delta\right\} \tag{1.6}$$

where $g$ is the gradient of $f$ at the current point $x$, $H$ is the Hessian matrix (the symmetric matrix of second derivatives), $D$ is a diagonal scaling matrix, $\Delta$ is a positive scalar, and $\|.\|$ is the 2-norm and T stands for vector transposition.

By finding the minima of this function in the vicinity of the initial estimate the first iteration of the optimization process is found. This iteration is used as the initial estimate for second iteration. When the difference between values of objective function in two consecutive iterations is less than the specified value, the last iteration is considered local minimum. All points of potential minima of the objective function are found initially. The smallest element of this group is the global minimum of the objective function in a prescribed area under consideration. The solution is based on MATLAB implementation of a modified method "Trust-Region Methods for Nonlinear Minimization".

In Fig. 1.11 is shown sample velocity profile, optimized for two time constraints on reference route. The results indicate that the time limit is the most important parameter of optimization.

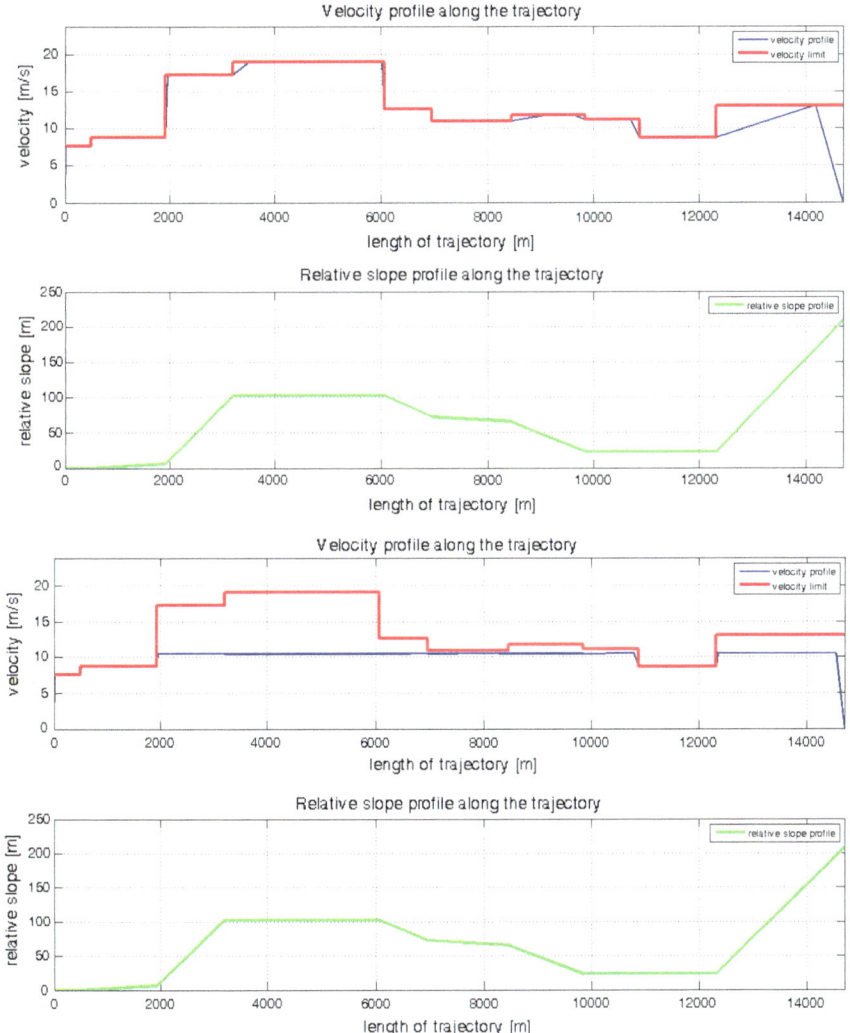

**Fig. 1.11** Optimal velocity profile for ref. route with maximum time 1300 s (**a**) and 1500 s (**b**)

## 1.6  Close Loop Control Design

The real drive always differs from pre-optimized profile due to the surrounding traffic and vehicle model differences. That's why the on-board real-time controller must be used to adapt the actual torque to achieve optimal behavior. It is based on gain scheduled, model based predictive controller, respecting constraints [6]. The linearized models around selected operating points in the grid

$$\Delta \dot{x} = A\Delta x + B\Delta u$$
$$\Delta y = C\Delta x + D\Delta u \tag{1.7}$$

with inputs: cabin external temperature difference request [°C], vehicle load [kg], wind speed [m s$^{-1}$], normalized torque [−], road slope [%] and outputs: vehicle acceleration [m/s$^2$], HV battery voltage [V], HV battery current [A], derivative of SOC [s$^{-1}$], velocity [m s$^{-1}$] were used. The linearization of the full simulation model provides relatively high order linear system with many states. Thus the model reduction is used to obtain control design model of acceptable size. Unfortunately, the resulting linear model has generally non-zero D matrix, which is not suitable for a standard MPC formulation. Thus integral methods [6] based on fast artificial states

$$\dot{\underline{Y}} = \frac{1}{T}(-\underline{Y} + \underline{C}.\underline{X} + \underline{D}.\underline{U}), \quad T \ll \frac{1}{|Re(\lambda_{max})|} \tag{1.8}$$

have been used. For each operating point the linear model was derived and used for control design.

The MPC design algorithm finds optimal controller with respect to quadratic criteria. However, there are additional design parameters, like sampling period, length of prediction horizon, maximum controlled variable rate etc.

The ultimate objective of this control design is to reduce energy consumption. Pre-optimization of the vehicle profile [7] has shown that maximum time of travel affects maximum energy consumption considerably. That's why multi-objective criteria optimization based on genetic algorithms was used with two conflicting criteria to determine best feedback controller settings

$$J_1 = \int_0^t 1 \, dt \tag{1.9}$$

$$J_2 = E_{MPC} = \int_0^t P_{mot} \, dt = \int_0^t \omega_{mot} \cdot M_{mot} \, dt \tag{1.10}$$

The individual (controller) was tested on shortened route (500 m) for acceptable optimization time. The genetic algorithms used 150 individuals in 5 generations.

**Fig. 1.12** Pareto set for conflicting criteria

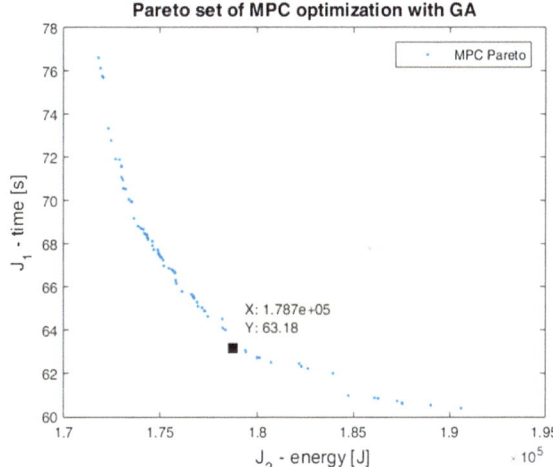

The result of the optimization is organized into Pareto set (Fig. 1.12), which shows possible suitable region of combinations $J_1$ and $J_2$ criteria.

## 1.7   Conclusions

The combined control strategy for electric vehicle has been demonstrated, together with preliminary results. The comparison has shown that control design based on accurate model predictive control provides significant improvements of control behaviour over traditional techniques. The non-linear control based on MPC gain scheduling and MPC Pareto optimization are demonstrated to be effective tools for control approach for energy savings. The final decision is left to a driver, who sets optimization constraints and may override of them during a trip.

**Acknowledgements** This research has been realized using the support of EU FP 7 Project No. 608756, Integration and Management of Performance and Road Efficiency of Electric Vehicle Electronics and using the support of The Ministry of Education, Youth and Sports program NPU I (LO), project# LO1311 Development of Vehicle Centre of Sustainable Mobility. This support is gratefully acknowledged.

## References

1. MINETT, Claire F et al (2011) Eco-routing: comparing the fuel consumption of different routes between an origin and destination using field test speed profiles and synthetic speed profiles. In: IEEE forum on integrated and sustainable transportation system (FISTS) 2011, p 32–39
2. Badin F, Le Berr F, Castel G, Dabadie JC, Briki H, Degeilh P, Pasquier M (2015) Energy efficiency evaluation of a plug-in hybrid vehicle under European procedure, Worldwide harmonized procedure and actual use; EVS28 KINTEX, Korea, 3–6 May 2015

3. Here. https://company.here.com
4. Maroteaux D, Le Guen D, Chauvelier E (2015) Development of a fuel economy and $CO_2$ simulation platform for hybrid electric vehicles, application to renault EOLAB prototype. RENAULT SAS, SAE International
5. Moré JJ, Sorensen DC (1983) Computing a trust region step. SIAM J Sci Stat Comput 3:553–572
6. Steinbauer P, et al (2017) E-vehicle predictive control for range extension. Advances in intelligent systems and computing
7. Steinbauer P et al (2016) Dynamic optimization of the e-vehicle route profile. No. 2016–01-0156. SAE technical paper, 2016
8. Seborg DE (2010) Process dynamics and control. John Wiley and Sons, USA
9. Biegler LT (2007) An overview of simultaneous strategies for dynamic optimization. Chem Eng Process 46(11):1043–1053

# Semi-autonomous Driving Based on Optimized Speed Profile

Sebastiaan van Aalst, Boulaid Boulkroune, Shilp Dixit,
Stephanie Grubmüller, Jasper De Smet, Koen Sannen
and Wouter De Nijs

## 2.1 Introduction

Electric vehicles (EVs) are becoming more and more a viable alternative to conventional internal combustion engine vehicles (ICEVs). Recent advances in battery capacity and cost made EVs more popular than ever. Next to their higher energy efficiency and lower emissions, EVs have some other remarkable advantages compared with ICEVs, such as their faster driving/braking torque response, and the possibility to have multiple motors enabling individual wheel control. Research to exploit these advantages is being actively conducted [1, 2].

Despite the recent advances in battery technology, the range of an EV per charge is still shorter than that of a conventional ICEV. One way to increase the driving range is by improving the efficiency of the drivetrain and energy storage. Another approach is to modify the driving performance by considering and taking advantage of all possible environmental information resulting in more efficient driving. A suitable way to reduce energy consumption can be devised by calculating an energy optimal speed profile for the chosen route by considering road and vehicle characteristics and real-time traffic and weather information [3, 4]. Such a system could operate in two modes: advisory mode, in which the energy optimal speed profile is displayed to the driver as a suggestion, and speed control mode, in which a longitudinal control system is employed such that the car automatically follows the optimal speed profile. This chapter will focus on the design and implementation

S. van Aalst (✉) · B. Boulkroune · S. Dixit · J. De Smet · K. Sannen · W. De Nijs
Flanders Make, Oude Diestersebaan 133, 3920 Lommel, Belgium
e-mail: Sebastiaan.vanaalst@flandersmake.be

S. Grubmüller
VIRTUAL VEHICLE Research Center, Inffeldgasse 21a, 8010 Graz, Austria
e-mail: Stephanie.Grubmueller@v2c2.at

© The Author(s) 2017
D. Watzenig and B. Brandstätter (eds.), *Comprehensive Energy Management—
Eco Routing & Velocity Profiles*, Automotive Engineering: Simulation
and Validation Methods, DOI 10.1007/978-3-319-53165-6_2

aspects of the speed control mode. This mode is an extension of Adaptive Cruise Control (ACC) that is already available in many production vehicles. The history of ACC can be traced back to the 1960s [5] but it was first patented by General Motors in 1991 [6]. The interested reader is referred to the review paper of Xiao et al. [7] which provides an excellent overview of the development of ACC. While ACC is designed for tracking a constant driver-set speed and distance keeping, the speed control mode can enhance driving efficiency by tracking an energy optimal speed profile. It is interesting to note that this added ability does not come at the cost of major changes in the existing control system structure.

In general, the ACC implementation is done in an hierarchical manner with an upper and a lower level controller [8]. The upper level controller determines the required torque for tracking the speed profile and the lower level controls the actuators. The upper level controller is usually developed as a cascade control system consisting of two control loops; The inner-loop controller is the speed tracking controller. It modulates the torque demand in order to track the reference speed. The outer-loop controller is the distance tracking controller. It calculates the reference speed in order to keep a desired distance from a leading vehicle and introduces this new reference speed into the inner-loop. This chapter will focus on the design of the inner-loop controller; on enhancing it to track a highly dynamic optimal speed profile, as required for the speed control mode. The design of the outer-loop controller is not the topic of this chapter but it has been shown that conventional control techniques such as proportional-integral-derivative (PID) control and its modifications work very well for the outer-loop [9–11].

For the speed control mode, the inner-loop or speed tracking controller plays a vital role in optimal driving as its performance directly influences how well the vehicle tracks the optimal speed profile. For its performance specifications, it is necessary to specify that the steady state tracking error should be zero. Furthermore, it should provide the desired acceleration/jerk behavior to ensure the occupants comfort. In literature many control approaches have already been proposed to meet these objectives; In [12] the performance of a conventional PI controller is compared with that of a fuzzy and neuro-fuzzy controller for low speed applications. A controller based on a sliding mode technique is proposed in [13]. In [9, 10, 14] gain scheduling PI(D) control is proposed. In [14] the authors also propose an adaptive control scheme. A nonlinear control strategy based on a direct Lyapunov design is proposed in [15] and [16]. More advanced approaches based on optimal control theory are proposed in [9, 10]. As far as the commercially available ACC is concerned, their control system design is closely guarded by the individual companies for competitive advantage. Hence it is not possible to gain a lot of insight into the implementation details of the state-of-practice. However, some of the patents filed in the current century show that none of them combine the approach taken in the current research of precomputing an optimal speed profile and using this as reference for the upper controller.

The controllers presented in the literature review from the preceding paragraph have in common that they have been developed mainly with the aim of tracking a quasi-constant speed profile. Furthermore, they are typically difficult to tune for the desired acceleration/jerk behavior for good ride comfort, which is very important when tracking a highly dynamic speed profile. To overcome these shortcomings, this chapter presents the design and hardware implementation of two model-based control approaches for the inner-loop that directly incorporate these design objectives in their control design: a novel exponentially stabilizing gain scheduling proportional integral controller (gs-PI), and a state-of-the-art offset-free explicit model predictive controller with preview (e-MPC). To the knowledge of the authors, these control approaches have not yet been applied for longitudinal control of EVs.

The proposed gs-PI controller follows a nonlinear Lyapunov-based control design that ensures exponential convergence stability. It has been designed to have the desired convergence rate as its design parameter. Given this value, an offline optimization-based procedure finds the scheduled gains that realize exponential convergence of the vehicle speed to the reference speed with the desired convergence rate. The exponential convergence behavior results in smooth speed transients with low jerk and thus good ride comfort. The e-MPC controller is based on optimal control theory in which the control action is obtained by solving, at each sampling instant, a finite-horizon open-loop optimal control problem, using the current state of the system as the initial state [17, 18]. This control approach can use the predefined optimal speed profile as preview data, and allows to directly specify the desired trade-off between tracking accuracy and ride comfort in the control problem formulation. In this work the explicit-MPC formulation is employed to shift the online optimization of MPC to an offline optimization problem.

The remainder of this chapter is structured as follows. Section 2.2 gives a brief overview of how to calculate an energy optimal speed profile for a given route. Section 2.3 discusses the control design for the gs-PI and e-MPC controller. Section 2.4 presents the results of an experimental validation on a 4 wheel drive rolling road with an EV test vehicle. Finally, Sect. 2.5 presents some concluding remarks.

## 2.2  Energy Optimal Speed Profile

The driving range of an EV can be increased by considering and taking advantage of all possible environmental information resulting in more efficient driving. This can be achieved by calculating an energy optimal speed profile and employing a longitudinal control system to automatically follow this speed profile. This speed profile is the solution of an optimization problem over the entire route chosen by the driver at the beginning of the journey. For the current research the optimization considers road and vehicle characteristics and cloud information such as route, weather, and traffic information. To solve this optimization problem and respect the speed limits the dynamic programming algorithm DPM [19] provided by the Swiss Federal Institute of Technology Zürich was used. It generates a grid containing all possible solution states resolved over distance and the desired resolution of the state space. Caused by a very large grid, the computation of the optimal speed profile requires high processing power, which is not available on the test vehicle of this research (equipped with an Infineon AURIX™ platform [20]). To overcome this problem, it was decided to deploy a high-performance computing platform to calculate the speed profile, namely a state-of-the-art equipped personal computer. As a consequence, the reference speed is provided by an off-board system, see Fig. 2.1.

The speed limits are obtained by fusing the speed constraints given by government, weather information, traffic flow information and road curvature. As a result, the general optimization problem can be written as:

$$
\begin{aligned}
&\underset{\mu(s)}{\text{Minimize}} \quad J(\mu(s)) \\
&\text{Subject to:} \quad \dot{x}(s) = F(x(s), \mu(s), s), \\
&\qquad\qquad\quad x(0) = x_0, \\
&\qquad\qquad\quad x(s_e) \in \left[x_e^{\min}, x_e^{\max}\right], \\
&\qquad\qquad\quad x(s) \in X(s) \subset \Re^n, \\
&\qquad\qquad\quad \mu(s) \in M(s) \subset \Re^n
\end{aligned}
\tag{2.1}
$$

where $F(\cdot)$ represents the longitudinal dynamic vehicle model. This model accounts for the effect of rolling resistance, aerodynamic drag and road grade, and also includes simple models for the transmission, electric motor/generator, battery and supercapacitors (if implemented) to describe their induced power losses. When applying the DPM, a quasi-static vehicle model is expected [21]. It assumes that the velocity for a very small distance is constant. For that, the track is resolved over distance and not over time. The function $J(\cdot)$ represents the optimization goal. It contains the power demand, which is expected to be minimized. The speed limits are implemented as soft constraints in $J(\cdot)$. For a detailed explanation of the DPM the interested reader is referred to Chap. 3 of this book.

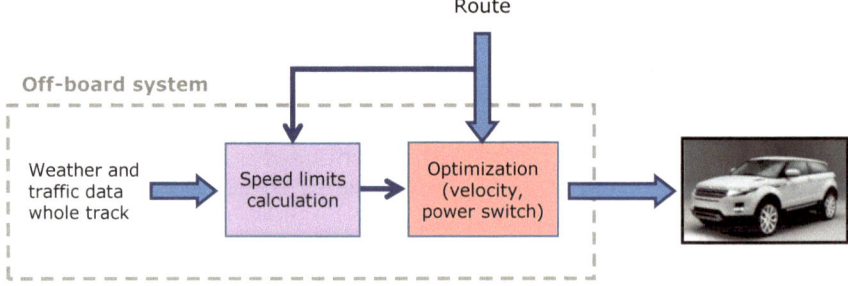

**Fig. 2.1** Block scheme of the off-board system for calculating the energy optimal speed profile

## 2.3  Controller Design

The longitudinal control system architecture for the EV is designed to be hierarchical, with an upper and lower level controller, as shown in Fig. 2.2. The upper-level controller determines the desired torque for tracking the speed profile and the lower-level controls the actuators. In the performance specifications for the upper-level controller, it is necessary to specify that the steady state tracking error should be zero. Other performance specifications concern the occupants comfort and are related to the acceleration/jerk profile. As far as the lower-level is concerned, it is assumed that its dynamics are much faster than that requested by the upper-level controller. This is a valid assumption as the electric motors that provide the desired torque (positive and negative) have a very fast torque response. Because of this, the lower-level control does not need to be considered in the control design of the upper-level, i.e. it is assumed that the torque requested by the upper-level controller can be delivered instantaneously.

The upper-level controller is developed as a cascade control system consisting of two control loops, as shown in Fig. 2.3; The inner-loop controller is the speed tracking controller. It modulates the torque demand in order to track the reference speed. This torque demand is input to the lower-level control. The outer-loop controller is the distance tracking controller. If a leading vehicle is detected, it modifies the reference speed in order to keep a desired distance from the leading vehicle and introduces this new reference speed into the inner-loop controller.

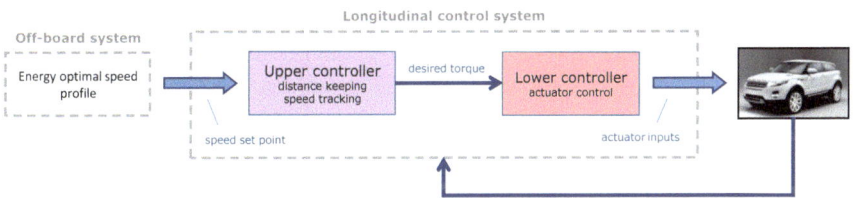

**Fig. 2.2**  Architecture of the longitudinal control system

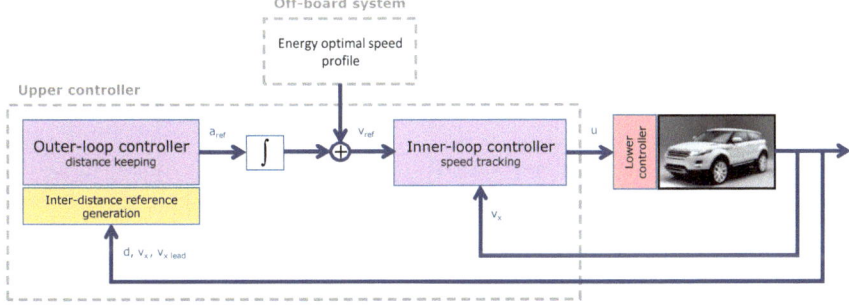

**Fig. 2.3**  Architecture of the longitudinal control system. Upper controller in detail

Several policies exist for the desired distance between two vehicles, such as a constant distance or constant-time headway [7]. The design of the outer-loop controller is not the topic of this chapter but it has been shown that conventional control techniques such as PID control and its modifications work very well for the outer-loop [9–11].

This section focuses on the control design for the inner-loop which can be made independently from the outer-loop. For its performance specifications, it is necessary to specify that the steady state tracking error should be zero. Furthermore, it should be easily tunable to obtain the desired acceleration/jerk profile to ensure the occupants comfort. To this end, this section presents two model-based control approaches for the inner-loop that directly incorporate these design objectives in their control design:

- Exponentially stabilizing gain scheduling proportional-integral control (gs-PI)
- Offset-free explicit model predictive control with preview (e-MPC).

### 2.3.1 Gain Scheduling Proportional-Integral Control

The gain scheduling proportional-integral-derivative (gs-PID) controller can be considered as the industry standard for control automation [22]. Its main advantage is that linear control design techniques can be applied to design a set of local linear controllers, whose control signals will be interpolated based on the current operating point of the system. The downside of this approach is that stability and performance are only guaranteed locally and global stability is not guaranteed for fast variations of the operating condition [23]. In order to address these issues, a nonlinear Lyapunov-based control design method has been applied to design a gain-scheduling PI controller for the inner-loop that ensures exponential stability. The general control architecture is shown in Fig. 2.4. The developed controller has the desired convergence rate as its design parameter. Given this value, it provides the scheduled gains that realize exponential convergence of the vehicle speed to the reference speed with the desired convergence rate. The exponential convergence leads to smooth speed transients with low jerk and thus good ride comfort.

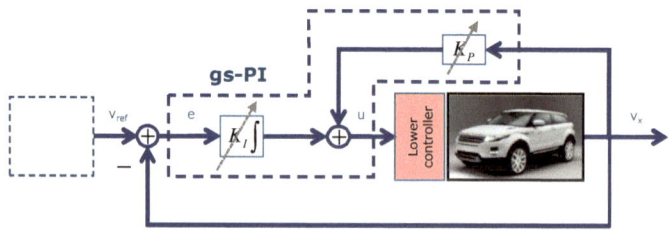

**Fig. 2.4** Block scheme of the proposed gain scheduling PI controller

The speed tracking controller is based on the following reduced longitudinal dynamics model of the vehicle:

$$\dot{v}_x = \frac{1}{m}\left(\frac{T}{r_{wh}} - \frac{1}{2}\rho C_d A_{fr} v_x^2\right) \tag{2.2}$$

in which $v_x$ is the vehicle speed, $T$ the vehicle-level torque demand, $m$ the vehicle mass including the apparent mass associated with the drivetrain components, $r_{wh}$ the wheel radius, $\rho$ the density of air, $C_d$ the aerodynamic drag coefficient, and $A_{fr}$ the frontal surface area of the vehicle. Notice that the rolling resistance is not included in this model. As it is almost a constant, its effect will be compensated for by the integral action of the controller. The same assumption is made for the other unknown external disturbances, e.g. due to road slope and wind speed.

Let us define $x = v_x$ as state, $u = T$ as control input, $y = v_x$ as measured output, and $\gamma(x) = v_x$ as scheduling variable with following upper and lower bound:

$$\gamma_{lb} := \gamma\left(v_x^{min}\right) \le \gamma(x) \le \gamma_{ub} := \gamma\left(v_x^{max}\right) \tag{2.3}$$

As a result, the longitudinal dynamics model can be expressed in quasi linear parameter-varying (LPV) form as follows:

$$\begin{cases} \dot{x} = (\theta_1 A_1 + \theta_2 A_2)x + Bu \\ y = Cx \end{cases} \tag{2.4}$$

with constant modes $A_1 = A(\gamma_{lb})$ and $A_2 = A(\gamma_{ub})$, $A(\gamma) = -1/(2m)\rho C_d A_{fr}\gamma(x)$, $B = 1/(mr_{wh})$ and $C = 1$. The time varying parameters $\theta_1$ and $\theta_2$ are given by:

$$\theta_1 = \frac{\gamma_{ub} - \gamma(x)}{\gamma_{ub} - \gamma_{lb}} \quad \text{and} \quad \theta_2 = \frac{\gamma(x) - \gamma_{lb}}{\gamma_{ub} - \gamma_{lb}}$$
$$\text{where } \theta_1 + \theta_2 = 1 \quad \text{and} \quad \theta_1 \ge 0, \theta_2 \ge 0 \tag{2.5}$$

The tracking problem is treated by means of the following gain scheduled control law:

$$u = K_P(\theta)y + K_I(\theta)\int_0^t (y_{ref} - y)dt \tag{2.6}$$

where $K_P(\theta) = \theta_1 K_P^1 + \theta_2 K_P^2$ and $K_I(\theta) = \theta_1 K_I^1 + \theta_2 K_I^2$ the interpolating gains, with $K_P^i \in \Re^{1\times 1}$ and $K_I^i \in \Re^{1\times 1}$ for $i = 1, 2$, and $y_{ref}$ the reference vehicle speed. Under this control law, the system can be equivalently represented by the following augmented closed-loop system associated to the state $\xi = [x\ z]^T$ where $z = \int_0^t (y_{ref} - y)dt$:

$$\dot{\xi} = \begin{bmatrix} A(\theta) + BK_P(\theta) & K_I(\theta) \\ -C & 0 \end{bmatrix} \xi + \begin{bmatrix} 0 \\ y_{\text{ref}} \end{bmatrix}$$

$$\xi(0) = \begin{bmatrix} x(0) \\ 0 \end{bmatrix} \tag{2.7}$$

The following theorem shows the sufficient linear matrix inequality (LMI) conditions for the existence of the gains $K_P(\theta)$ and $K_I(\theta)$ that ensure exponential stability of the previous closed-loop system.

**Theorem 1** *There exists a stabilizing PI-type control such that the augmented closed-loop system (2.7) is exponentially stable with a prescribed constant convergence rate $r > 0$ (that is, $\lim_{t \to +\infty} e^{rt} x = 0$); if there exist a symmetric matrix $P \in \Re^{2 \times 2}$, and constant matrices $G_P \in \Re^{1 \times 1}$, $G_I \in \Re^{1 \times 1}$, $Y_P^1 \in \Re^{1 \times 1}$, $Y_P^2 \in \Re^{1 \times 1}$ and $Y_I^1 \in \Re^{1 \times 1}$, $Y_I^2 \in \Re^{1 \times 1}$, such that the following condition holds true (involving the prescribed scalar $s > 0$ as parameter design):*

$$\begin{bmatrix} -2sP & M_i^T + P + (r - s)G^T \\ M_i + P + (r - s)G & G + G^T \end{bmatrix} < 0 \tag{2.8}$$

*where $G = diag(G_P, G_I)$ and*

$$M_i = \begin{bmatrix} A_i G_P + B Y_P^i C & Y_I^i \\ -C G_P & 0 \end{bmatrix} \quad for \ i = 1, 2 \tag{2.9}$$

*If this LMI condition is feasible, then the gains $K_P^i$ and $K_I^i$ can be computed as:*

$$K_P^i = Y_P^i G_P^{-1} \ and \ K_I^i = Y_I^i G_I^{-1}, \quad i = 1, 2 \tag{2.10}$$

*Proof* The exponential stability of the closed-loop system can be proved by considering a quadratic Lyapunov function $V = \xi^T P \xi$ where $P$ satisfies the above LMI condition. With some mathematical manipulations it can be shown that $V$ satisfies $\dot{V}(t) < -2rV(t)$. Then by Grönwalls lemma $\dot{V}(t) < \dot{V}(0)e^{-2rt}$, that is $\xi^T P \xi < \xi^T(0)P\xi(0)e^{-2rt}$ which implies $\lambda_{\min}(P)\xi^T \xi < \xi^T(0)P\xi(0)e^{-2rt}$ (where $\lambda_{\min}(P)$ is the smallest eigenvalue of P). This in turn implies $\xi^T \xi < \frac{\xi^T(0)P\xi(0)}{\lambda_{\min}(P)}e^{-2rt}$ if $\lambda_{\min}(P) > 0$ or equivalently $P$ is a positive definite matrix. The fact that $P > 0$ results from $s > 0$ and the fact that $-2sP$ is the upper diagonal block of the negative definite matrix in the LMI condition. Hence, from these facts it can be concluded that the trajectory of the closed-loop system will exponentially converge to the reference with the prescribed constant convergence rate $r > 0$.

To ensure the exponential stability of the closed-loop system, the derivative of the quadratic Lyapunov function $V = \xi^T P \xi$ should satisfy $\dot{V}(t) < -2rV(t)$ which is equivalent to:

$$\left(\bar{A}_i^T + rI + sI\right)P + P(\bar{A}_i + rI + sI) - 2sP < 0 \quad \text{for } i = 1, 2 \tag{2.11}$$

where $\bar{A}_i$ is the transition matrix of closed-loop system (2.7) and given by:

$$\bar{A}_i = \begin{bmatrix} A_i + BK_P^i C & K_I^i \\ -C & 0 \end{bmatrix} \quad \text{for } i = 1, 2 \tag{2.12}$$

From [24] the previous relation can be expressed as:

$$\begin{bmatrix} -2sP & \left(\bar{A}_i^T + (r+s)I\right)G^T + P \\ G\left(\bar{A}_i^T + (r+s)I\right) + P & G + G^T \end{bmatrix} < 0 \tag{2.13}$$

To retrieve the LMIs in the theorem, it suffices to take $Y_P^i = K_P^i G_P$ and $Y_I^i = K_I^i G_I$, $i = 1, 2$, where $G = \text{diag}(G_P, G_I)$. This ends the proof.

The controller gains can be obtained by solving the LMI conditions (2.8)–(2.10). The SDPT3-4.0 solver from Yalmip toolbox [25] can be used for this.

The proposed control approach has some major benefits compared to conventional (gain scheduling) PI(D) control. They can be summarized as follows:

- The gains are obtained from an off-line model-based procedure.
- Exponential stability of the closed-loop system is ensured.
- The controlled output converges exponentially to the desired reference. This results in smooth speed transients with low jerk and thus good ride comfort.
- Easily extendable to include model parameter uncertainties in the controller design as well as disturbance rejection.

### 2.3.2 Model Predictive Control with Preview

The second type of control that is utilized for the inner-loop is model predictive control. This is a form of control in which the current control action is obtained by solving, at each sampling instant, a finite-horizon open-loop optimal control problem, using the current state of the system as the initial state [18]. The optimization yields the optimal control sequence and the first control in this sequence is applied to the system. Figure 2.5 shows a schematic representation of this closed-loop process.

For the current use-case, MPC has some important advantages over traditional PID-type control approaches:

- The control objective, that is, the desired trade-off between tracking accuracy and ride comfort, can be directly specified in the control problem formulation.
- MPC can explicitly handle constraints, such as the actuator constraints.

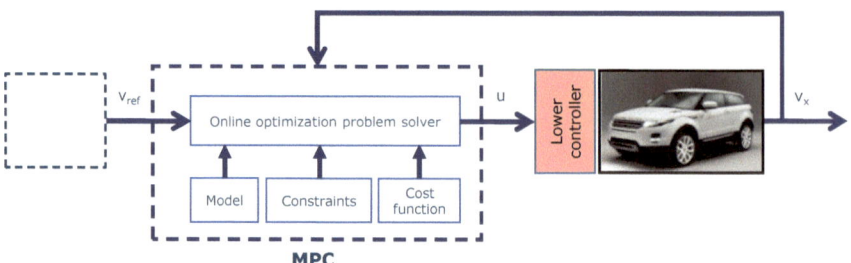

**Fig. 2.5** Block scheme of a Model Predictive Controller (MPC) [30]

- The inherent predictive nature of MPC allows it to anticipate known future events. This is particularly useful for the current use-case as the optimal speed reference is known in advance.

The MPC controller for speed tracking is based on a point-mass model of the vehicle. The model only considers the driving and braking torque of the vehicle, $T$. All other effects, such as aerodynamic drag, rolling resistance, road slope, etc. are treated as unknown disturbances. In order to obtain offset-free control in the presence of these disturbances, two common approaches exist; The first approach augments the system with an integrator state that integrates the error between predicted output and reference trajectory. Penalizing this integrator state in the objective function drives the offset to zero. The second approach augments the system with a disturbance model (e.g. $\dot{F}_d = 0$) to capture the mismatch between measured and predicted outputs [26]. An observer designed for the augmented system is used to simultaneously estimate this disturbance and the system state. The estimates are then used to initialize the MPC control problem at each sampling instant. In this work the second approach is applied due to the availability of the observer and previous experiences of the authors with this formulation. The resulting discrete-time model for the controller is shown in Eq. (2.14). An input increment formulation is used so that the optimization routine can reduce jerk by moderating the rate of change of torque input.

$$x_{(k+1)} = Ax_{(k)} + B\Delta u_{(k)}$$

$$\begin{bmatrix} v_{x(k+1)} \\ F_{d(k+1)} \\ T_{(k+1)} \end{bmatrix} = \begin{bmatrix} 1 & \frac{\Delta t}{m} & \frac{\Delta t}{mr_{wh}} \\ 0 & 1 & 0 \\ 0 & 0 & 1 \end{bmatrix} \begin{bmatrix} v_{x(k)} \\ F_{d(k)} \\ T_{(k)} \end{bmatrix} + \begin{bmatrix} \frac{\Delta t}{mr_{wh}} \\ 0 \\ 1 \end{bmatrix} \Delta T_{(k)} \qquad (2.14)$$

where $x_{(k)} \in \Re^{n_x}$ and $\Delta u_{(k)} \in \Re^{n_u}$ with $n_x = 3$ and $n_u = 1$. $\Delta t$ is the time step size, and subscript $k+i$ refers to time sample $k+i$.

To achieve good reference tracking performance with smooth control input, the objective function shown below is used for the MPC optimization routine. As evident from the equation, this function computes the objective cost of tracking error over prediction horizon $(H_p)$ and input increment over control horizon $(H_c)$.

$$J(x, \Delta u) = \sum_{i=0}^{H_p} \left\| x_{(k+i)} - x_{\text{ref}(k+i)} \right\|_Q^2 + \sum_{i=0}^{H_c-1} \left\| \Delta u_{(k+i)} \right\|_R^2 \tag{2.15}$$

This objective function in conjunction with the system model (2.14) and subject to constraints on states and inputs forms a quadratic programming (QP) problem expressed in a standard format below:

$$\begin{aligned} &\underset{w}{\text{Minimize}} && \tfrac{1}{2} w^T H w + f^T w \\ &\text{Subject to} && G w \le b, \\ &&& G_{\text{eq}} w = b_{\text{eq}}, \\ &&& w_{\min} \le w \le w_{\max} \end{aligned} \tag{2.16}$$

$$\begin{aligned} \text{where} \quad & w = \left[ x_{(0)}, \Delta u_{(0)}, \ldots x_{(H_c-1)}, \Delta u_{(H_c-1)}, x_{(H_c)}, \ldots x_{(H_p)} \right]^T, \\ & f = -1 \left( w_{\text{ref}}^T H \right)^T \end{aligned}$$

in which $w$ is the optimization vector, $w_{\text{ref}}$ is the stacked vector containing state and input reference trajectory over the prediction horizon, $H$ the symmetric and positive semi-definite Hessian matrix, and $f$ the gradient vector. $G$ and $b$ define the inequality constraints, $G_{\text{eq}}$ and $b_{\text{eq}}$ the equality constraints, and $w_{\min}$ and $w_{\max}$ the lower and upper bounds respectively.

The MPC algorithm solves this optimization problem at every sampling instant. This requires fast hardware and reliable software solutions which can be a major challenge for real-time implementation on the vehicle. To circumvent this problem, the online QP problem is converted into a multi-parametric QP (mpQP) problem using the YALMIP and MPT3 toolbox in MATLAB to develop an explicit-MPC (e-MPC) controller [25, 27]. This toolbox requires that the parameters to the mpQP problem are passed as a single input argument. To this end, the initial state, initial input and reference trajectory are combined into a single parametric variable $x_p$:

$$x_p = \begin{bmatrix} x_{(k+0)} \\ w_{\text{ref}} \end{bmatrix} \tag{2.17}$$

The cost function and constraints are then written in terms of the parameter $x_p$ as:

$$\begin{aligned} &\underset{w}{\text{Minimize}} && \tfrac{1}{2} w^T H w - x_{p(n_x+1:end)}^T H w \\ &\text{Subject to} && G w \le b, \\ &&& G_{\text{eq}} w = b_{\text{eq}}, \\ &&& w_{\min} \le w \le w_{\max}, \\ &&& w_{\min} \le x_{p(n_x+1:end)} \le w_{\max} \end{aligned} \tag{2.18}$$

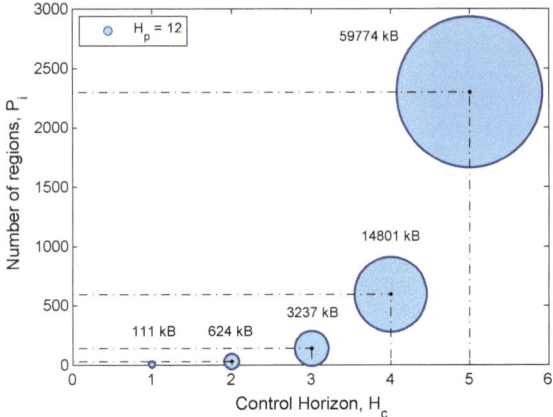

**Fig. 2.6** Visual representation of the number of regions $(P_i)$ and size of lookup table as a function of control horizon $(H_c)$. The area of the circle represents the relative size of the lookup table on memory (Values generated for the current use-case.)

The mpQP optimization (2.18) is performed offline and the solution to this problem gives rise to a piece-wise affine control law:

$$u = F_i x_p + G_i, \text{ if } x_p \in P_i$$
$$P_i := \{x_p \in \Re^n | E_i x_p \le k_i\} \tag{2.19}$$
$$\text{where, } n = (H_p + 2) \cdot n_x + H_c \cdot n_u$$

Thus, the solution to the mpQP problem converts the MPC control to a linear state feedback law (2.19) where $P_i$ are the various polytopic regions, $F_i$ and $G_i$ are linear gains, and $E_i$ and $k_i$ define the vertices and edges of the polytopic regions. The bulk of computation is now performed offline and the solution can be implemented on real-time hardware in the form of a lookup table. The complexity of the mpQP problem and of the polytopic solution depends in decreasing order on the number of constraints, control horizon $(H_c)$ and number of parameters, i.e. prediction horizon $(H_p)$ and state dimension $(n_x)$ [28]. The number of regions $(P_i)$ increases exponentially with an increase in control horizon $(H_c)$ and consequently this increases the computational requirements of parsing the lookup table. The evolution of number of regions and size of lookup table as a function of control horizon $(H_c)$ can be observed in Fig. 2.6. Thus, for successful implementation on a hardware setup a balance between size of lookup table, computational requirements and control horizon must be achieved.

## 2.4   Experimental Validation

This section discusses the experimental validation. First, the test vehicle is introduced. Next, the experimental results are presented and discussed.

### 2.4.1   Test Vehicle

The test vehicle is the electric vehicle demonstrator of the European Union-funded projects E-VECTOORC [1] and iCOMPOSE [2], Fig. 2.7. This vehicle has four individual electric drivetrains with switched reluctance on-board motors, connected to the wheels through single-speed gearboxes, constant velocity joints and half-shafts. The vehicle has a traction battery (600 V, 9 kWh) designed to handle high current peaks in both traction and regenerative mode. Figure 2.8 shows the hardware architecture of the vehicle. The vehicle is equipped with a dSPACE AutoBox platform to control the electric drives and friction brakes. The presented controllers for longitudinal control are implemented on this platform. The torque set points for the electric drives are transmitted via controller area network (CAN) bus 1. The set points for the friction brakes are transmitted via CAN bus 2. The energy optimal speed profile is stored on an Infineon AURIX$^{TM}$ [20] multi-core platform and communicated via CAN bus 4 to the dSPACE vehicle controller. This information is also transmitted to the human-machine interface (HMI) such that the speed profile can also be displayed to the driver (advisory mode). The AURIX platform is connected to the internet to access real-time traffic and weather information.

**Fig. 2.7** The 4 wheel drive electric vehicle demonstrator during a test on the rolling road at Flanders Make

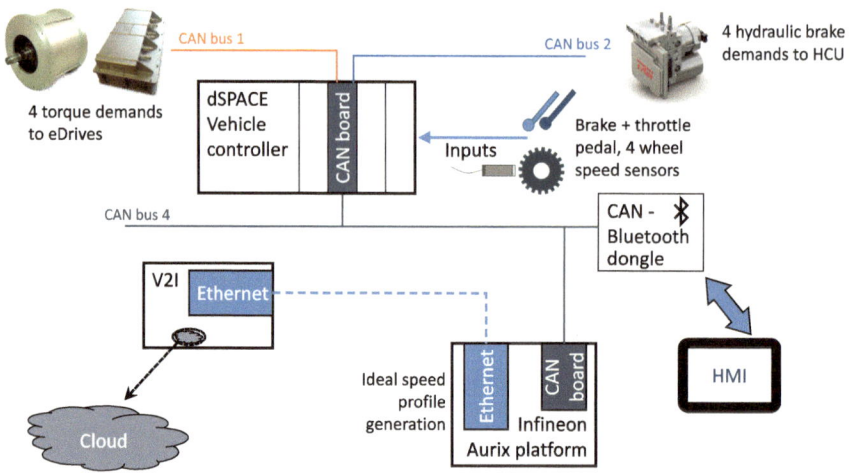

**Fig. 2.8** The hardware architecture of the 4 wheel drive electric vehicle demonstrator (*CAN* Controller Area Network, *HCU* Hydraulic Control Unit, *HMI* Human-Machine Interface, *V2I* Vehicle-to-Infrastructure)

## 2.4.2 Experimental Results

The presented controllers were validated on a 4 wheel drive rolling road, Fig. 2.7. The rolling road was configured at road load simulation such that it emulates the inertial load of the vehicle and the aerodynamic drag and rolling resistance forces (based on a coast down test). The reference speed profile was chosen to contain various target speeds and accelerations corresponding to normal driving. For each controller two tests were performed with a different tuning:

- gs-PI: The tuning implies choosing the value of the convergence rate parameter *r*. The higher this value, the faster the convergence but also the higher the jerk. For this test the following values were used:

  - Controller 1: $r = 1.5$, for which following scheduled gains are obtained:

  $$K_P^1 = 2.3652 \times 10^3, \quad K_P^2 = 2.3747 \times 10^3$$

  $$K_I^1 = 1.7816 \times 10^3, \quad K_I^2 = 1.7816 \times 10^3$$

- Controller 2: $r = 2.0$, for which following scheduled gains are obtained:

**Table 2.1** Number of regions and occupied memory for implemented e-MPC controller

| $H_p$ | $H_c$ | Number of regions | Size (kB) |
|---|---|---|---|
| 8 | 1 | 6 | 61 |
| 15 | 1 | 6 | 157 |

$$K_P^1 = 3.1566 \times 10^3, \quad K_P^2 = 3.1662 \times 10^3$$

$$K_I^1 = 3.1670 \times 10^3, \quad K_I^2 = 3.1670 \times 10^3$$

Consequently, controller 2 should convergence more quickly than controller 1.

- e-MPC: The tuning implies choosing the length of the prediction and control horizon and setting the cost function weightings that define the desired trade-off between tracking accuracy and torque demand smoothness. For this test the following values were used:

  - Controller 1: $H_p = 15$ and $H_c = 1$, with following cost function weightings:

$$Q_{v_x} = 1 \times 10^6, \quad Q_{F_d} = Q_T = 0$$

$$R = 1 \times 10^{-5}$$

  - Controller 2: $H_p = 8$ and $H_c = 1$, with following cost function weightings:

$$Q_{v_x} = 2.5 \times 10^6, \quad Q_{F_d} = Q_T = 0$$

$$R = 1 \times 10^{-5}$$

Additional information about the number of regions and occupied memory is given in Table 2.1. Because of the higher penalty on speed error and an aggressive controller due to the shorter prediction horizon, one can expect controller 2 to provide better tracking performance than controller 1.

Note that the control horizon $H_c$ is kept equal to one for both e-MPC controllers. A simulation study showed that increasing $H_c$ allows for more aggressive control with better tracking performance. Too aggressive control, however, is not desired for the current use-case as this negatively impacts the ride comfort. Furthermore, increasing $H_c$ strongly increases the complexity of the control problem and the size of the lookup table, as was explained in Sect. 2.3.2. The choice of $H_c$ thus depends

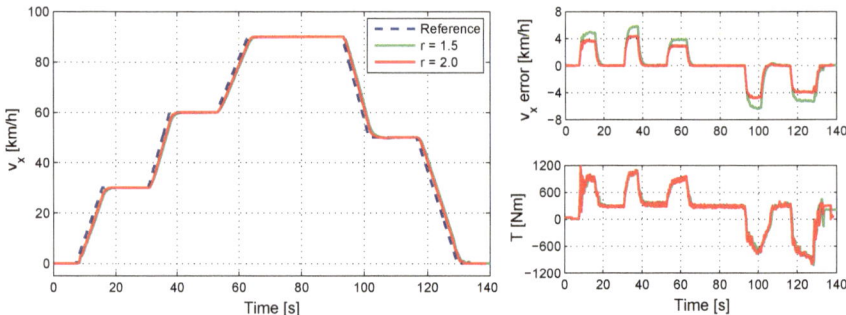

**Fig. 2.9** Experimental result: exponentially stabilizing PI control for r = 1.5 and r = 2.0. *Left* reference speed versus actual speed, *top right* speed tracking error, *bottom right* torque demand

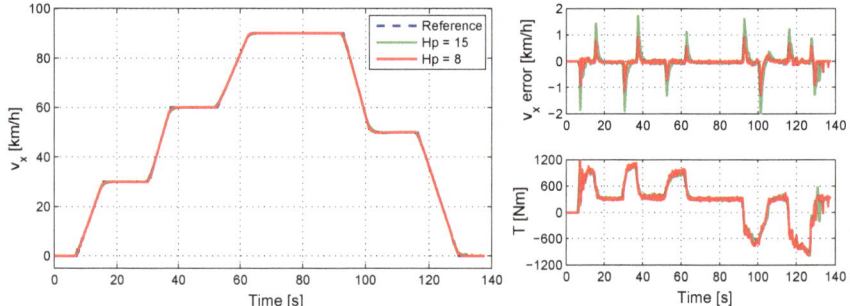

**Fig. 2.10** Experimental result: explicit model predictive control for $H_p = 15$, $H_c = 1$, and $H_p = 8$, $H_c = 1$. *Left* reference speed versus actual speed, *top right* speed tracking error, *bottom right* torque demand

on a compromise between complexity and computational requirements, tracking performance, and ride comfort. For the current use-case a control horizon of one turned out to be sufficient; it results in relatively small lookup-tables, and allows to obtain the desired trade-off between tracking performance and ride comfort.

Figures 2.9 and 2.10 show the results for the gs-PI and e-MPC controller, respectively. The left figure depicts the vehicle's speed with respect to the target speed. The top right figure shows the speed error and the bottom right figure the torque demand. Figure 2.11 is a zoom of Figs. 2.9 and 2.10. The results show very promising behavior: both controllers provide a smooth speed response and torque demand without oscillations. Both controllers also realize offset-free tracking in the presence of unmodeled disturbances (coming from the rolling road). The gs-PI controller achieves this through its integral action, the e-MPC controller through the disturbance model and observer. The control performance is further evaluated by means of two performance indicators; the root mean square (RMS) of vehicle speed tracking error to evaluate tracking performance, and the infinity norm of jerk

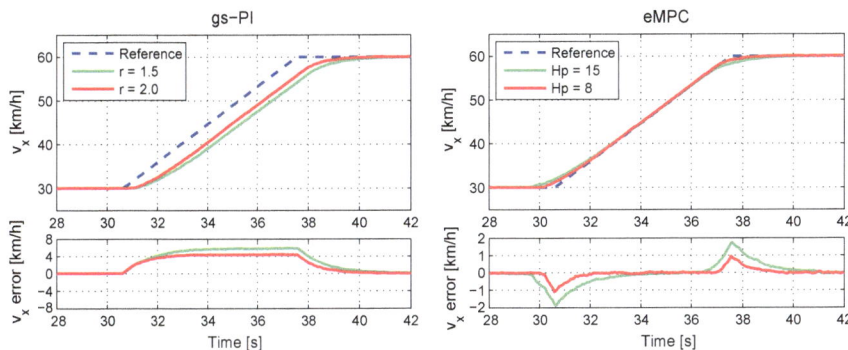

**Fig. 2.11** Zoomed view, *left* gs-PI, *right* eMPC. *Top* reference speed versus actual speed, *bottom* speed tracking error

**Table 2.2** Performance indicators for the gs-PI and e-MPC controller

|  | gs-PI | | e-MPC | |
|---|---|---|---|---|
|  | $r = 1.5$ | $r = 2.0$ | $H_P = 15$ | $H_P = 8$ |
| RMS of tracking error [km/h] | 2.81 | 2.17 | 0.40 | 0.18 |
| Infinity norm of jerk [g/s] | 0.094 | 0.096 | 0.093 | 0.104 |

(second derivative of vehicle speed) to evaluate ride comfort. For the latter, a limit of 0.3 g/s should typically be respected for ride comfort [29]. The results of this analysis are presented in Table 2.2. A third common performance indicator concerns the amount of acceleration overshoot. The presented controllers, however, did not show any acceleration overshoot and therefore this indicator is not added to the table.

The performance indicators show that the gs-PI and e-MPC controller combine accurate speed tracking with good ride comfort. Performance-wise, the e-MPC controller outperforms the gs-PI, mainly because it has the added benefit of preview. This allows it to anticipate future changes in reference speed by accelerating (decelerating) the vehicle ahead of time, see Fig. 2.11. In this way, the tracking error for speed transients is further reduced while maintaining good ride comfort. This comes at the cost of a more complex control with several tuning parameters. The gs-PI controller, on the other hand, is a more simple controller that allows for easy implementation and tuning and yet provides very good control performance. Its main benefit is that the Lyapunov-based design ensures exponential convergence stability, and results in only one tuning parameter; the convergence rate parameter $r$. Its exponential convergence behavior represents an excellent trade-off between tracking accuracy and ride comfort. This is a major benefit for practical implementation.

## 2.5 Conclusion

This chapter presented the design and hardware implementation of two model-based control approaches for longitudinal control of an electric vehicle: a novel exponentially stabilizing gain scheduling proportional-integral controller (gs-PI), and a state-of-the-art offset-free explicit model predictive controller with preview (e-MPC). These controllers directly incorporate the most important design objectives in their control design: tracking performance and ride comfort. The gs-PI controller follows a nonlinear Lyapunov-based control design that ensures exponential convergence stability. Given the desired convergence rate, the scheduled gains are obtained from an offline optimization-based procedure. The e-MPC controller is based on optimal control theory. The desired trade-off between tracking accuracy and ride comfort can be directly specified in the control problem formulation. The explicit MPC formulation was employed to successfully implement an optimal controller in the form of lookup-tables on a real-time hardware setup.

The proposed controllers were validated experimentally with an electric vehicle demonstrator on a 4 wheel drive rolling road for a realistic driving scenario. The test results and performance indicators show the effectiveness of both control approaches; they provide accurate speed tracking with good ride comfort and realize offset-free tracking in the presence of unmodeled disturbances. Performance-wise, the e-MPC controller outperforms the gs-PI, mainly due to the added benefit of preview. This comes at the cost of a more complex control with several tuning parameters. The gs-PI controller, on the other hand, is a more simple controller that allows for easy implementation and tuning and yet provides very good control performance. Its exponential convergence behavior represents an excellent trade-off between tracking accuracy and ride comfort.

**Acknowledgements** This work was supported by the European Union Seventh Framework Programme FP7/2007-2013 under the iCOMPOSE project (grant agreement no. 608897). The authors also acknowledge the financial support of the COMET K2—Competence Centres for Excellent Technologies Programme of the Austrian Federal Ministry for Transport, Innovation and Technology (BMVIT), the Austrian Federal Ministry of Science, Research and Economy (BMWFW), the Austrian Research Promotion Agency (FFG), the Province of Styria and the Styrian Business Promotion Agency (SFG).

## References

1. www.e-vectoorc.eu. Last accessed on 28th June 2016
2. www.i-compose.eu/iCompose. Last accessed on 28th June 2016
3. Kuriyama M, Yamamoto S, Miyatake M (2010) Theoretical study on eco-driving technique for an electric vehicle with dynamic programming. International conference on electrical machines and systems (ICEMS), 2026–2030
4. Sciarretta A, De Nunzio G, Ojeda LL (2015) Optimal ecodriving control: energy-efficient driving of road vehicles as an optimal control problem. IEEE Control Syst 35(5):71–90
5. Diamond H, Lawrence W (1966) The development of an automatically controlled highway system. University of Michigan, Transportation Research Institutes's Report UMR1266

6. Chundrlik WJ, Labuhn PI (1991) Adaptive cruise system. US Patent 5,014,200
7. Xiao L, Gao F (2010) A comprehensive review of the development of adaptive cruise control systems. Veh Syst Dyn Int J Veh Mech Mobility 48(10):1167–1192
8. Rajamani R (2006) Vehicle dynamics and control. Springer
9. Shakouri P, Ordys A, Laila DS, Askari M (2011) Adaptive cruise control system: comparing gain-scheduling PI and LQ controllers. In: Proceedings of the 18th IFAC world congress, Milano (Italy)
10. Shakouri P, Czeczot J, Ordys A (2014) Simulation validation of three nonlinear model-based controllers in the adaptive cruise control system. J Intell Robot Syst 80(2)
11. Martinez JJ, Canudas-de-Wit C (2007) A safe longitudinal control for adaptive cruise control and stop-and-go scenarios. IEEE Trans Control Syst Technol 15(2):246–258
12. Milanés V, Villagrá J, Pérez J, González C (2012) Low-speed longitudinal controllers for mass-produced cars: a comparative study. IEEE Trans Industr Electron 59(1):620–628
13. Nouveliere L, Mammar S (2007) Experimental vehicle longitudinal control using a second order sliding mode technique. Control Eng Pract 15:943–954
14. Ioannou P, Xu Z, Eckert S, Clemens D, Sieja T (1993) Intelligent cruise control: theory and experiment. In: Proceedings of the 32nd conference on decision and control, San Antonio, Texas
15. Attia R, Orjuela R, Basset M (2014) Combined longitudinal and lateral control for automated vehicle guidance. Veh Syst Dyn 52(2):261–279 (Taylor & Francis: STM, Behavioural Science and Public Health Titles)
16. El Majdoub K, Giri F, Ouadi H, Dugard L, Chaoui FZ (2012) Vehicle longitudinal motion modeling for nonlinear control. Control Eng Pract 20:69–81
17. Rawlings JB, Mayne DQ (2009) Model predictive control: theory and design. Nob Hill Publishing
18. Mayne DQ, Rawlings JB, Rao CV, Scokaert POM (2000) Constrained model predictive control: stability and optimality. Automatica 36(6):789–814
19. Sundstrom O, Guzzella L (2009) A generic dynamic programming matlab function. IEEE control applications (CCA) & intelligent control (ISIC), 1625–1630 (2009)
20. www.infineon.com. Last accessed on 30th June 2016
21. Guzzella L, Sciaretta A (2007) Vehicle propulsion systems. Springer, Berlin
22. Aström K, Hägglund T (2006) Advanced PID controllers. ISA—The Instrumentation, Systems and Automation Society, Research Triangle Park, NC
23. Shamma JS (1992) Gain scheduling: potential hazards and possible remedies. IEEE Control Syst 12(3):101–107
24. Song L, Yang J (2011) An improved approach to robust stability analysis and controller synthesis for LPV systems. Int J Robust Nonlinear Control 21(13):1574–1586
25. Löfberg J (2004) YALMIP: a toolbox for modeling and optimization in matlab. In: Proceedings of the CACSD conference, Taipei, Taiwan
26. Borrelli F, Manfred M (2007) Offset free model predictive control. In: Proceedings of the 46th IEEE conference on decision and control, pp 1245–1250, New Orleans, LA, USA
27. Herceg M, Kvasnica M, Jones CN, Morari M (2013) Multi-parametric toolbox 3.0. In: Proceedings of the European control conference, pp 502–510, Zurich, Switzerland. http://control.ee.ethz.ch/∼mpt
28. Stephens MA, Manzie C, Good MC (2011) Explicit model predictive control for reference tracking on an industrial machine tool. IFAC Proc Volumes 44(1):14513–14518
29. Hoberock LL (1977) A survey of longitudinal acceleration comfort studies in ground transportation vehicles. J Dyn Syst Meas Contr 99(2):76–84
30. Abbaszadeh M, Solgi R (2014) Constrained nonlinear model predictive control of a polymerization process via evolutionary optimization. J Intell Learn Syst Appl 6(1):35–44

# The Design of Vehicle Speed Profile for Semi-autonomous Driving

Zdenek Herda, Pavel Nedoma and Jiri Plihal

## 3.1 Optimal Speed Profile Design

### 3.1.1 Basic Assumptions

The algorithm recommends optimal levels of throttle pedal, brake pedal and gear shift. This is based on a several factors. The decisive part is focused on the driving horizon rather than maximizing speed in the next route section. In this horizon the key point is selected (decisive point) which presents the largest decrease of speed. Given this point we optimize the change of speed in the closest time frame. The algorithm decides whether it is necessary to change parameters from the previous step based on the change of speed in that previous step. If it is necessary the algorithm makes a decision with respect to the current driving mode.

Before initiating braking or accelerating, the algorithm verifies the track profile between the current location and the key point. It modifies parameters of the braking distance and optimal speed based on the inclination. The following action is decided after this modification. The action flow diagram is shown in Fig. 3.1 and described later in Sect. 3.1.3.

Z. Herda (✉) · P. Nedoma
SKODA AUTO a.s., tr. Vaclava Klementa 869, 293 60 Mlada Boleslav, Czech Republic
e-mail: Zdenek.Herda@skoda-auto.cz

J. Plihal
UTIA AV CR, v.v.i., Pod Vodarenskou vezi 4, 182 08 Praha 8, Czech Republic

© The Author(s) 2017
D. Watzenig and B. Brandstätter (eds.), *Comprehensive Energy Management—Eco Routing & Velocity Profiles*, Automotive Engineering: Simulation and Validation Methods, DOI 10.1007/978-3-319-53165-6_3

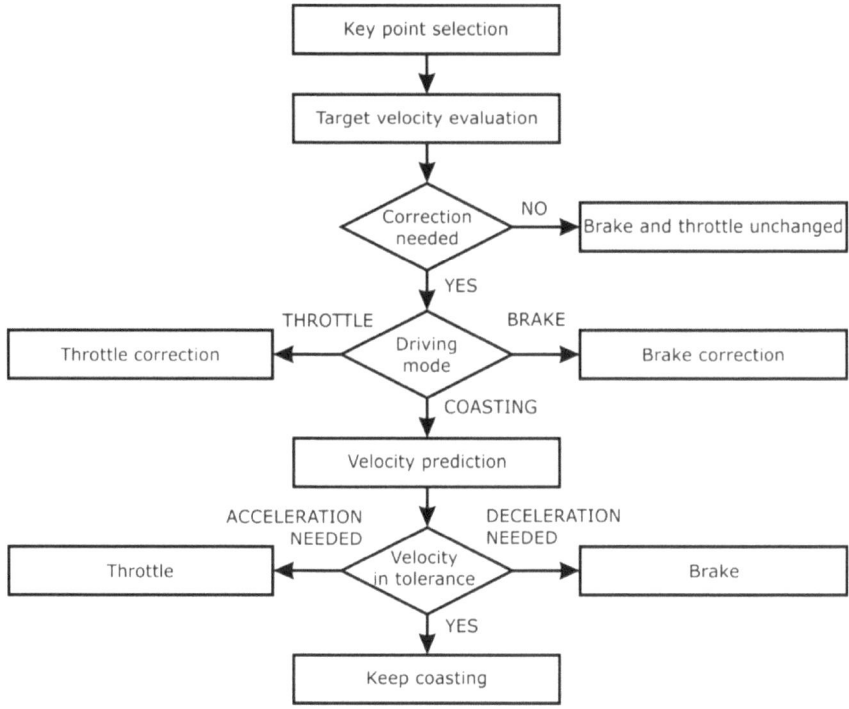

**Fig. 3.1** The scheme of brake and throttle motion

## 3.1.2 Target Velocity

Achieving the optimal speed in one step of the algorithm would not be effective and it could cause repeated corrections of the vehicle dynamics. As a basic preposition the algorithm takes respect to the part of the track in the distance $H$ (hundreds of meters). This enables the assignment of milestone to points on the track with respect to the discretization step of the road for example 20 m. The current location is then marked as 1. Variables are then indexed by these milestones, i.e., optimal speed $v_i^{opt}$, distance $x_i$, etc.

The point which represents the largest decrease of speed in the discretization step is selected as the key point with respect to the traffic density and changed legal limit. The key point is based on the topology of the road and road features. In case all points have assigned a larger speed than the actual vehicle's speed $v$ then the maximum decrease can be interpreted as the minimum increase. The mathematical representation follows:

$$r_i = \frac{v - v_i^{opt}}{x_i - x}, \tag{3.1}$$

$$k = \operatorname{argmax}_{i \in H}(r_i). \tag{3.2}$$

The goal of the algorithm is to achieve the speed $v_k^{opt}$ in the following path of the vehicle. With respect to the theoretical discretization of vehicle's movement, e.g. $\Delta t = 0.2$ [s], we can determine the traveled distance as $\Delta s = \Delta t * v/3.6$ [m]. Given the ratio between this distance and the distance to the key point the algorithm modifies the target speed change:

$$\Delta v_{NEXT} = (v - v_k^{opt}) \frac{\Delta s}{x_k - x}. \tag{3.3}$$

### 3.1.3   Vehicle's Movement Dynamic Correction

The algorithm works iteratively. It modifies value of speed $v$, throttle $p$ and brake $b$ from the previous step ($v_{LAST}, b_{LAST}, p_{LAST}$) in every time step instead of computing new values again every time. The change of speed in the last step is given by

$$\Delta v_{LAST} = v - v_{LAST}. \tag{3.4}$$

The planned speed change (3.3) and the last speed change (3.4) are compared to decide the necessity of intervention. For this comparison coefficient $C$ was defined which is basically the size of hysteresis.

$$\Delta v_{KOR} = \Delta v_{NEXT} - \Delta v_{LAST}. \tag{3.5}$$

If $|\Delta v_{KOR}| < C$, it is not necessary to intervene. Brake $b$ and throttle $p$ remains unchanged: $b = b_{LAST}$ and throttle $p = p_{LAST}$. The next step of the algorithm is based on the current driving mode—brake, throttle and coasting.

(a)  Brake:

In this mode the vehicle deceleration is controlled. If $\Delta v_{KOR} \leq -C$, it is necessary to apply more deceleration than the vehicle is currently doing. In opposite if $\Delta v_{KOR} > C$, it is necessary to apply less deceleration. The correction of braking force is in both cases based on the following formula:

$$b = b_{LAST} - \beta \Delta v_{KOR}. \tag{3.6}$$

The coefficient $\beta$ modifies the magnitude of the braking force change with respect to the natural change of speed in the comfort driving. When the value of $b$ decreases below a certain level $b_{min}$ (*based on the adhesion limit and vehicle assistant system activation*) the process of braking is terminated ($b = 0$).

(b)  Throttle:

In this mode the acceleration of the vehicle is controlled. The principle is similar to brake mode. If the value of $\Delta v_{KOR} \leq -C$, it is necessary to accelerate more than the current acceleration is. In opposite if $\Delta v_{KOR} > C$, the required acceleration is lower. The correction of throttle is in both cases based on the following formula:

$$p = p_{LAST} - \gamma \Delta v_{KOR}. \tag{3.7}$$

The coefficient $\gamma$ modifies the magnitude of the throttle position change with respect to the comfort driving. When the value of $p$ decreases below a certain level $p_{min}$ the acceleration process is terminated ($p = 0$).

(c)  Coasting:

During coasting mode the algorithm predicts vehicle velocity in coasting mode and considers if the predicted velocity fits into the defined limits like maximal deviation form speed limit and safety limit in curves. If the velocity fits the coasting mode remains unchanged. If not the coasting is canceled and algorithm switches the state into the throttle or brake based on need for acceleration or deceleration.

### 3.1.4  Shifting Gears

The most important fact for gear shifting is engine revolutions and load, but also the inclination of the road had to be considered. The need for shifting with electric motor is based on engine efficiency which depends on revolutions and load. The efficiency map will be presented later in Sect. 3.2.3. The recommended area of revolutions for each gear can be defined by 4 key points (see Fig. 3.2): minimal acceptable revolutions A, minimal revolutions for shifting down B, minimal revolutions for shifting up C, maximal acceptable revolutions D.

This approach breaks down the need for gear shifting into five basic intervals. The behavior for all intervals follows:

(1)  $\langle 0, A \rangle$—Downshift.
(2)  $\langle B, C \rangle$—No shifting required.
(3)  $\langle D, D+ \rangle$—Upshift.
(4)  $(A, B)$—Algorithm considers downshift. This is done only if one of the following conditions occurs:

**Fig. 3.2**  Key points for gear shifting

- The gas pedal is pressed further than in the previous step ($p_{LAST} < p$). In this case the gear remains the same.
- Large uphill climbing is not expected and simultaneously the acceleration is not active ($p = 0$). In this case the current shift is removed coasting-down is active. It is canceled in the moment when ($p > 0$). In this moment the previous shift is reengaged.

(5) (C, D)—Algorithm considers upshift. It occurs when the gas pedal is partially released ($p_{LAST} > p$).

The iterative system of changing action inputs of gas and brakes introduces the feeling of real driving into the model. It considers wider conditions such as inclination and rolling resistance. It allows choosing the system reaction more precisely and thus reduces the amount of corrections. Also, by setting the toleration area, it is possible to modify the amount of interventions and thus regulate the fluency of driving. The system of changing gears includes the possibility of coasting-down, when the vehicle uses the declining road or redundant speed to shift to the neutral gear. This leads to energy savings. The transition between the state where brakes are active and the state where the gas pedal is active is done by values $b = p = 0$, which prevents from pressing both pedals at once.

The proposed algorithm does not allow exceeding the maximum allowed speed which leads to the safe driving. Simulated energy consumption with and without shifting will be shown later on Table 3.2.

### 3.1.5  Model Approach

In the next procedure was for experimental verification used the optimal speed profile algorithm modelled by the Bayes approach described by the equation:

$$P(Y|X_1X_2...X_n) = \frac{P(X_1X_2...X_n|Y)}{P(X_1X_2...X_n)}P(Y) \qquad (3.8)$$

The probability occurrence of dependent quantity $Y$ in a specific category is the same as the joint probability distributions of these quantities evaluated for this category and multiplied by the probability of this category. Consider that $X_1X_2...X_n$ are measured quantities and $Y$ is the probability occurrence of quantities $X_1X_2...X_n$. $Y$ is normalised by the joint distribution of measured quantities. The shape of the distribution function is not limited, its selection should respect the empirical distribution of measured data and the nature of estimated quantities.

The joint distribution $P(X_1X_2...X_n|Y)$ for each category of velocity and energy consumption priority may be extracted from the driving test measurements together with the distributions of each marginal distribution. This enables the evaluation of probability distribution of estimated quantity in a given category and either select the most likely category in a given time or focus on the probability of appearance in one of the risky categories.

For better illustration the progress of described algorithm is shown on Figs. 3.4, 3.5, 3.6 and 3.7 where the priority coefficient alpha represents gain function as a weighted sum of two contradicting criteria—travel time and energy consumption. Higher value means higher average velocity (shorter travel time) and lower value means lower energy consumption.

### 3.1.6  Outputs from the Vehicle Speed Profile Model

This chapter is concerned with the optimal speed package tool. This tool is created to find optimal driving profiles for a car driving along a known path from point A to point B with a certain gain function. This function can take various forms and in our case it provides weighted optimisation of two contradicting criteria: time consumption and energy consumption. The goal is to provide an optimal leading speed and acceleration/deceleration signals to maximise the gain (i.e. minimise the loss in terms of travelling time):

$$con = t * \left( c_{p0} * k \right) \tag{3.9}$$

where $k$ is a control input ([1,10] accelerating, [−1, −10] decelerating), $t$ is time spent driving with such control input and $c_{p0}$ is parameter. This parameter is fitted form data and applies according to the quality of the fit. Please note that even with no control input there is some consumption present.

The approach to achieve the above stated goal lies in the use of Bayesian Networks in the form of decision graphs. We profit from the decomposability property of this type of graph and Bellman's principle leading to the possibility of local solutions. The path is traversed backwards obtaining optimal profiles leading to the next point in each step. When the algorithm reaches the start of the track, the forward pass constructs the optimal solution profile.

The model was learned to be able to work with a different drive type (electric) and the use of the neutral gear was heavily revised. Herein is present the overview of the current status with two different drives at the end.

The most important aspects of the model are functions modelling the vehicle and its consumption. So far we are using functions which were fitted from observed experimental data. It is necessary to point out that these data are of small volume. Due to a previous discretisation step some intervals of values were not treated properly. This behaviour led to varied results for different discretisation steps while all remaining variables stayed the same. The most apparent problem was in the discretisation of the acceleration. This behaviour was removed by the fix. For a better illustration of the algorithm progress we present a comparison with its different versions in the form of plots.

Circuit around Mlada Boleslav around 38 km in length, consisting of highway, primary roads and local roads.

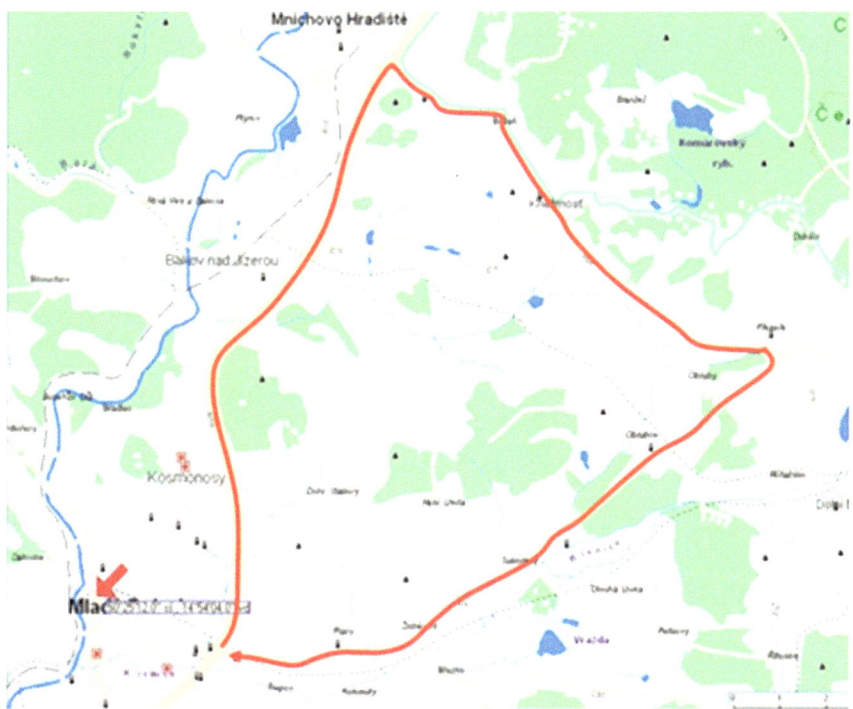

**Fig. 3.3**  Test course near Mlada Boleslav

The circuit was measured repeatedly, after verification and regularisation of metrics, correlations of the road surface parameters and dynamic features were assessed from the vehicle demonstrator Skoda eRapid FEV, see Fig. 3.3.

Results from testing the vehicle demonstrator Skoda eRapid on two testing tracks in the proximity of Mlada Boleslav, see Figs. 3.4, 3.5, 3.6 and 3.7.

The output from the final version of the algorithm without the coasting-down mode for different coefficient alpha shows Fig. 3.8. Theoretical average speed and energy consumption for these speed profiles are in Table 3.1.

## 3.2  Efficiency Driving Using Optimal Speed Profile and Coasting-Down

### 3.2.1  Coasting Mode Strategy

Consider the optimal speed profile with a balanced average speed and energy consumption due to selected coefficient *alpha* described in the previous text. It seems that the energy consumption cannot be further reduced, but there is still

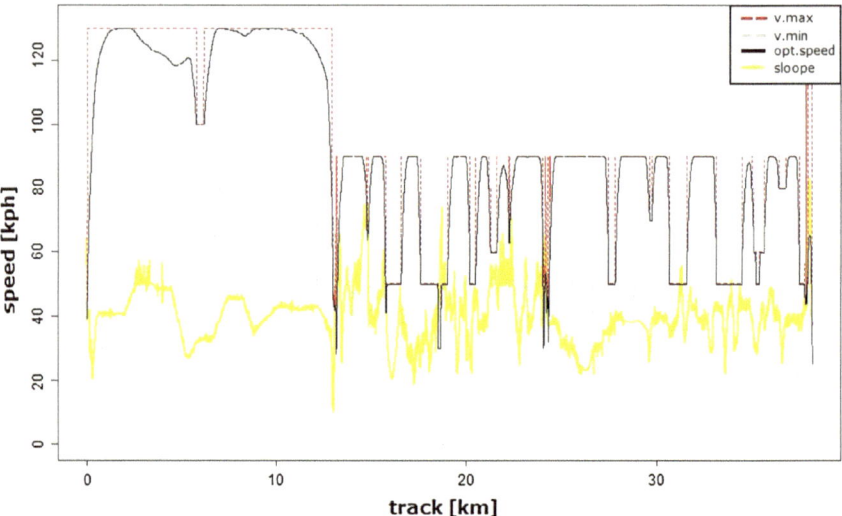

**Fig. 3.4** Model outputs, energy consumption 10.854 kWh, travel time 28.16 min, alpha 0.6, 38 km long track, without using coasting-down strategy

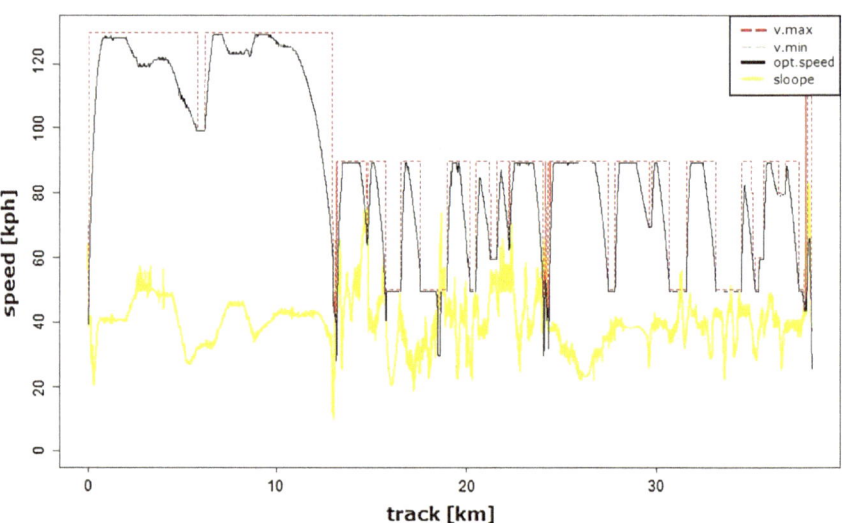

**Fig. 3.5** Model outputs, energy consumption 10.011 kWh, travel time 29.54 min, alpha 0.6, 38 km long track, with using coasting-down strategy

another option known as coasting. The original idea was to apply this mode directly to the optimal speed profile algorithm, but it was not suitable for a driver option to deactivate this mode during driving because the coasting behaviour affects the shape of the speed profile so it cannot be simply deactivated. In this case a different

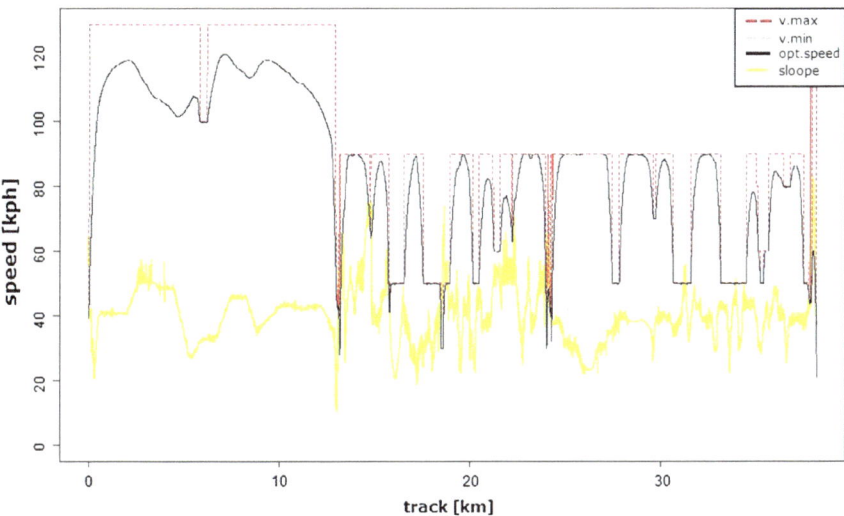

**Fig. 3.6** Model outputs, energy consumption 10.234 kWh, travel time 29.6 min, alpha 0.4, 38 km long track, without using coasting-down strategy

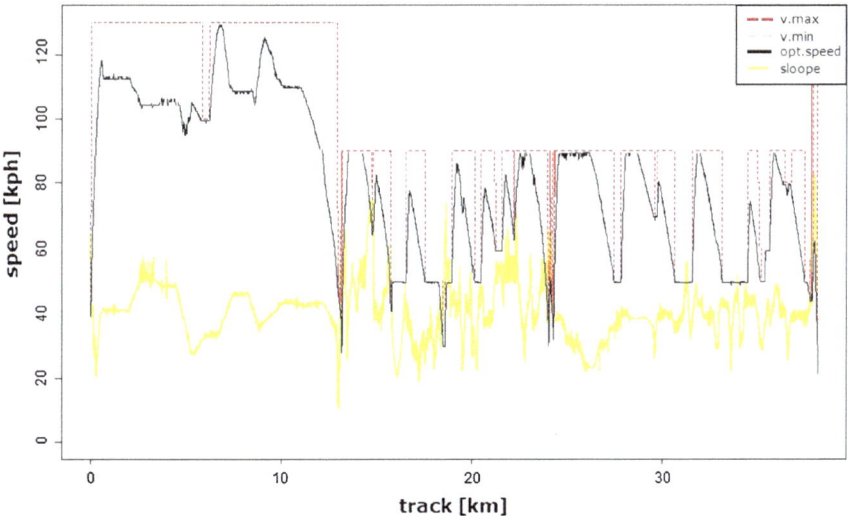

**Fig. 3.7** Model outputs, energy consumption 9.282 kWh, travel time 31.54 min, alpha 0.4, 38 km long track, with using coasting-down strategy

algorithm had to be designed only for this mode which can be applied on-line during driving.

Coasting can be applied to any car by using the neutral position of the gearbox. For our car demonstrator with an asynchronous electric engine it is possible to do in

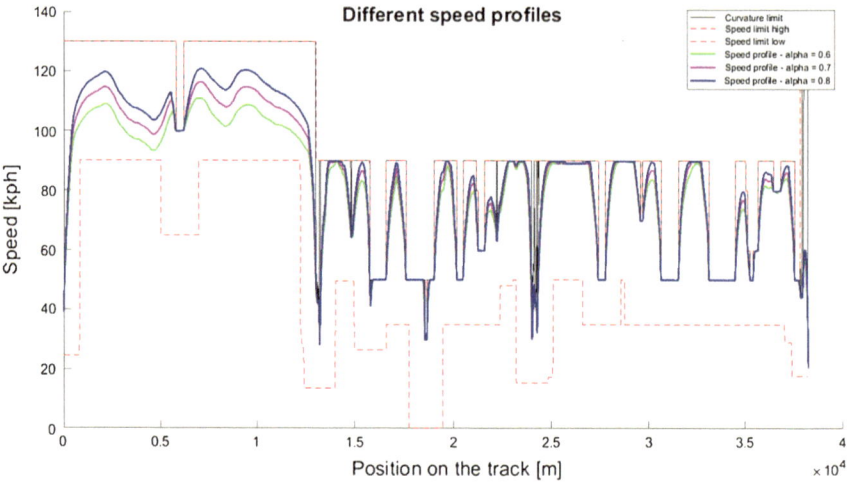

**Fig. 3.8** Model outputs for different coefficients alpha = 0.6, 0.7, 0.8

**Table 3.1** Theoretical energy consumption/regeneration and average speed of designed profiles

| Alpha | Energy per track [kWh] | | | Average speed [kph] |
|-------|------------------------|-----------|--------|---------------------|
|       | Consumed | Regenerated | Total |                     |
| 0.6   | 8.3964   | 0.4434      | 7.953  | 80.6228             |
| 0.7   | 8.6789   | 0.5088      | 8.1702 | 83.1274             |
| 0.8   | 8.9555   | 0.5765      | 8.3789 | 85.294              |

the electric way instead of using the neutral position. And how does it work? The principle is very simple. Accumulated kinetic energy of the car is used to continue driving with small deceleration which is more effective than keeping a higher velocity and then decelerating with recuperation. Of course coasting should only be applied at a suitable position on the track like highway exits, deceleration before curves and speed limits. The suitable areas are identified on-line using the model based prediction of the car velocity. The model structure is described in Sect. 3.2.3.

An example of coasting is shown on the speed profile calculated for coefficient *alpha* = 0.7 (see Fig. 3.9). The red line shows the speed profile without coasting and the blue line shows deceleration with coasting at suitable parts of the track. A detailed description of deceleration with coasting is shown on the highway exit (see Fig. 3.10). For a better comparison both pictures also show total consumed and regenerated energy (integration of actual energy consumption).

For the coasting algorithm must be defined some allowed offset from the original speed and minimal length of track during coasting. The exemplified coasting was calculated with the offset from the original velocity +1, −15 kph and minimal length 50 m. The coasting strategy in the example follows:

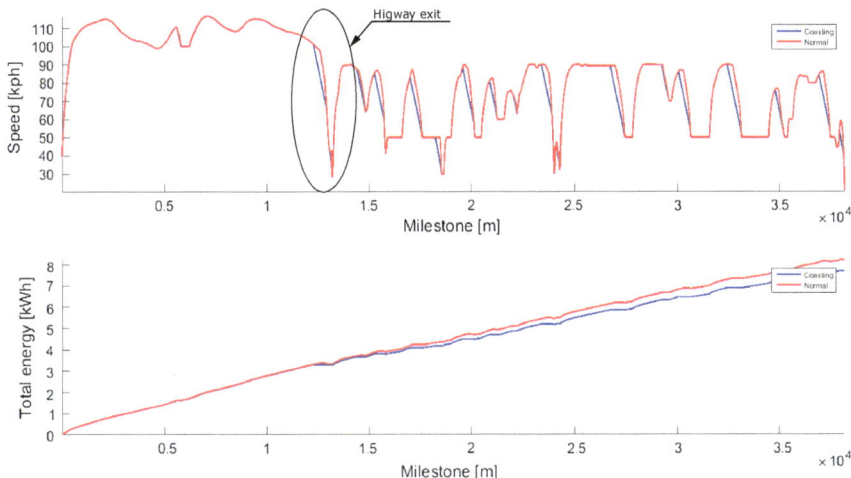

**Fig. 3.9** Optimal speed profile calculated for alpha = 0.7, *red*—speed profile without coasting, *blue*—deceleration using coasting

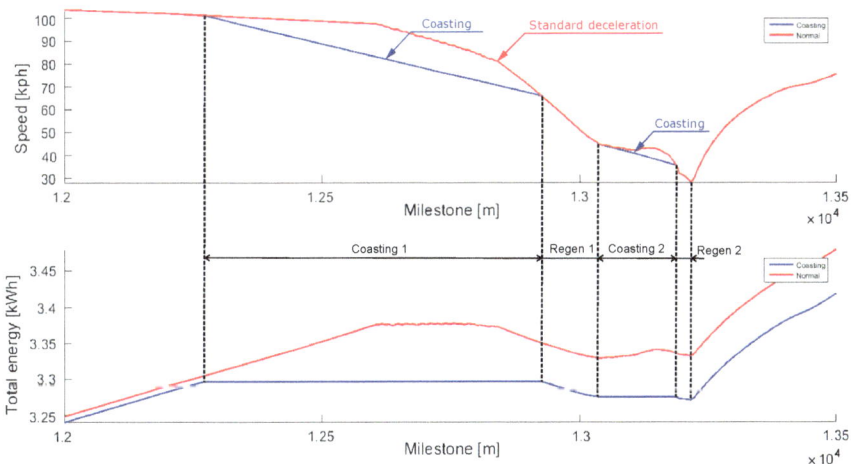

**Fig. 3.10** Detailed deceleration on highway exit using coasting

- Coasting 1—Apply coasting when described conditions are accomplished.
- Regen 1—Coasting deceleration is not enough, regeneration breaking must be used.
- Coasting 2—Conditions are accomplished again.
- Regen 2—Final regeneration breaking.

Coasting calculated for different offsets from the original speed (+1, −10, −15 and −20 kph) is a little different due to the original speed profile because not all

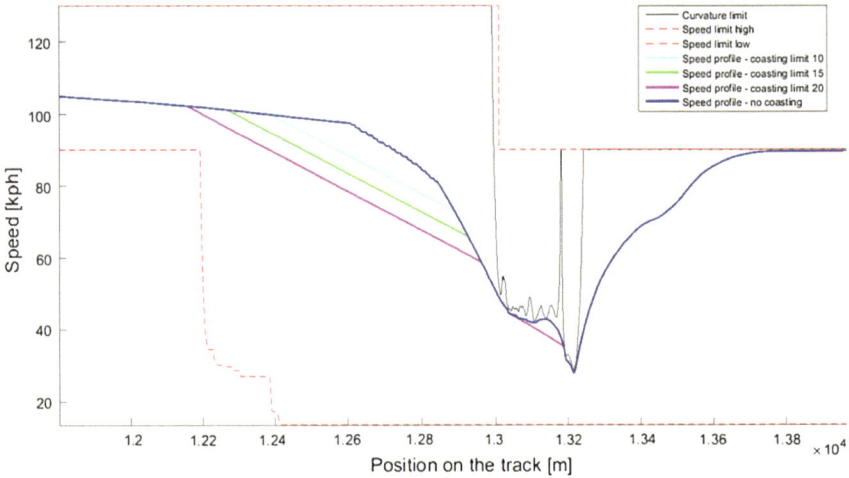

**Fig. 3.11** Coasting calculated for a different offset: −10, −15, −20 kph

decelerations are possible to use for the example maximal offset −20 kph. This example is also shown on highway exit (Fig. 3.11). The first deceleration using coasting is possible to make with all three defined offsets, but the last coasting deceleration no because of speed profile shape. It only allows one chance for coasting and all three calculated coastings are in cover so it looks like only one design.

It is obvious that coasting makes the average speed lower and it is necessary to find a good compromise between the average speed and saved energy during coasting. In the shown example the original speed profile average speed is 83.13 kph and with coasting 81.58 kph.

Simulated energy consumption for different speed profiles and different limits for coasting algorithm is shown on Table 3.2. The benefit of coasting in comparison with the same speed profile without coasting is almost 6%. An interesting comparison is also between the speed profile for alpha = 0.6 without coasting and alpha = 0.7 with coasting using tolerance +1, −15 kph. The average speed of profile 0.6 is 80.62 kph and the average speed of profile 0.7 is 81.58 kph. Although the average speed of profile 0.7 is higher the total consumed energy is 7.69 kWh which is less than 7.95 kWh for profile 0.6. This is the proof of higher efficiency of coasting in comparison with regeneration breaking.

### 3.2.2 Real Road Measurement Using Semiautonomous Driving with and Without Coasting Mode

As was already mentioned Semiautonomous driving could run in two different modes. The first is without coasting and second with coasting. The Optimal speed

**Table 3.2** Simulated results for different speed profiles and driving modes

| Driving mode | | | Energy per track [kWh] | | | | Average speed [kph] | Shifting benefit [%] | Coasting benefit [%] |
|---|---|---|---|---|---|---|---|---|---|
| Alpha | Shifting | Coasting | Tolerance | Consumed | Regenerated | Total | | | |
| 0.6 | on | off | – | 8.3964 | 0.4434 | 7.953 | 80.6228 | 1.3018 | – |
| | on | on | +1 −10 | 7.7364 | 0.1431 | 7.5934 | 79.7072 | 1.2998 | 4.5216 |
| | on | on | +1 −15 | 7.5768 | 0.0789 | 7.4979 | 79.181 | 1.2694 | 5.7224 |
| | on | on | +1 −20 | 7.5023 | 0.0488 | 7.4535 | 78.891 | 1.2311 | 6.2806 |
| | off | off | – | 8.4908 | 0.4329 | 8.0579 | 80.6228 | – | – |
| | off | on | +1 −10 | 7.8283 | 0.135 | 7.6934 | 79.7072 | – | 4.5235 |
| | off | on | +1 −15 | 7.6674 | 0.0732 | 7.5943 | 79.181 | – | 5.7534 |
| | off | on | +1 −20 | 7.5919 | 0.0455 | 7.5464 | 78.891 | – | 6.3478 |
| 0.7 | on | off | – | 8.6789 | 0.5088 | 8.1702 | 83.1274 | 1.3452 | – |
| | on | on | +1 −10 | 7.9993 | 0.192 | 7.8073 | 82.2357 | 1.3433 | 4.4418 |
| | on | on | +1 −15 | 7.7958 | 0.108 | 7.6878 | 81.5808 | 1.2942 | 5.9044 |
| | on | on | +1 −20 | 7.6865 | 0.0639 | 7.6226 | 81.1781 | 1.2604 | 6.7024 |
| | off | off | – | 8.7783 | 0.4966 | 8.2816 | 83.1274 | – | – |
| | off | on | +1 −10 | 8.0946 | 0.181 | 7.9136 | 82.2357 | – | 4.4436 |
| | off | on | +1 −15 | 7.8894 | 0.1008 | 7.7886 | 81.5808 | – | 5.9530 |
| | off | on | +1 −20 | 7.779 | 0.0591 | 7.7199 | 81.1781 | – | 6.7825 |
| 0.8 | on | off | – | 8.9555 | 0.5765 | 8.3789 | 85.294 | 1.2330 | – |
| | on | on | +1 −10 | 8.262 | 0.2453 | 8.0167 | 84.4485 | 1.2211 | 4.3228 |
| | on | on | +1 −15 | 8.0306 | 0.1475 | 7.8831 | 83.7536 | 1.1858 | 5.9172 |
| | on | on | +1 −20 | 7.8716 | 0.0834 | 7.7882 | 83.1585 | 1.1386 | 7.0499 |
| | off | off | – | 9.0462 | 0.5627 | 8.4835 | 85.294 | – | – |
| | off | on | +1 −10 | 8.3478 | 0.2319 | 8.1158 | 84.4485 | – | 4.3343 |
| | off | on | +1 −15 | 8.1151 | 0.1373 | 7.9777 | 83.7536 | – | 5.9622 |
| | off | on | +1 −20 | 7.9544 | 0.0765 | 7.8779 | 83.1585 | – | 7.1386 |

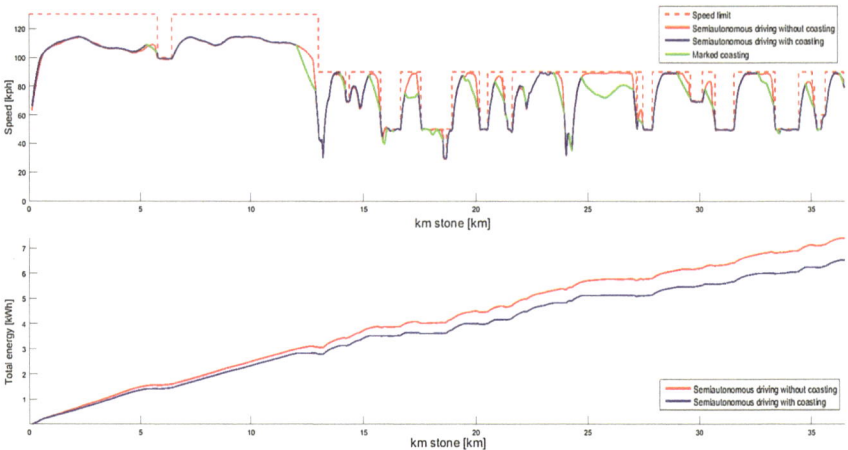

**Fig. 3.12** Semiautonomous driving with and without coasting

profile, which is an input to the system, is always designed without coasting and coasting algorithm is a part of the car speed controller so it can be simply activated or deactivated. This was done to see different energy consumption on the track in both modes. The coasting mode makes the average speed of the vehicle a little lower than the mode without coasting but it has the significant benefit of lower energy consumption. This benefit is higher than using regenerative breaking, because energy regeneration is not as effective as driving and due to two energy conversions.

The first conversion is from battery to kinetic and potential energy (driving) and the second conversion from kinetic and potential energy into battery (regeneration). Both conversions have some efficiency which means two energy losses. Coasting has in a principle much better efficiency, because there is only one energy conversion from battery to kinetic and potential energy.

The second conversion from kinetic and potential energy to battery is missing and accumulated kinetic and potential energy is used only to overcome car driving resistance. Comparison of both driving modes is shown in Fig. 3.12 and measured results like real average speed and energy consumption are listed in Table 3.3.

### 3.2.3 Simple Vehicle Energy Consumption Model

The goal was to find a simple vehicle model which describes energy consumption of the car and is suitable for on-line calculations in automotive control units. This model is based on the balance of forces in longitudinal direction of driving (Fig. 3.13) and torque balance in the centre of the front wheel (Fig. 3.14).

**Table 3.3**  Results of semiautonomous driving with and without coasting

Measured energy consumption for driving mode with and without coasting

| Driving mode | Average speed [km/h] | Energy per track [kWh] | | | Energy consumption [kWh/100 km] |
|---|---|---|---|---|---|
| | | Consumed | Regenerated | Total | |
| Without coasting | 77.31 | 7.9962 | 0.5815 | 7.4146 | 20.39 |
| With coasting | 75.54 | 6.5263 | 0.2482 | 6.5263 | 17.94 |
| Benefit of coasting [%] | | | | | 12.02 |

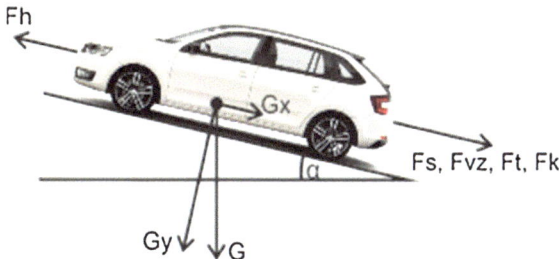

**Fig. 3.13**  Balance of forces in longitudinal direction

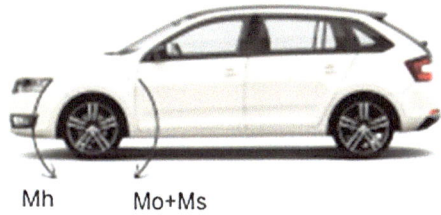

**Fig. 3.14**  Torque balance in the center of the front wheel

$$F_h = F_s + F_{vz} + F_v + G_x + F_k \qquad (3.10)$$

where

| | |
|---|---|
| $F_h$ | driving power |
| $F_s$ | inertia force |
| $F_{vz}$ | air resistance |
| $F_v$ | rolling resistance |
| $G_x$ | part of gravity force |
| $F_k$ | wheel resistance during cornering |

$$M_h = M_o + M_s \tag{3.11}$$

$$M_h = (F_s + F_{vz} + F_v + G_x + F_k) * R + M_s \tag{3.12}$$

$$M_h = \left( m * v' + \frac{1}{2} * v^2 * \rho * C_w A + m * g * \cos \alpha * \xi + m * g * \sin \alpha + F_k \right) *$$
$$* R + J_p * \omega_k' \tag{3.13}$$

where

| | |
|---|---|
| $M_h$ | driving torque |
| $M_o$ | resistance torque |
| $M_s$ | powertrain inertia torque |
| $R$ | wheel radius |
| $m$ | vehicle mass |
| $\rho$ | air density |
| $C_w A$ | product of drag coefficient $C_x$ and front surface of the car |
| $g$ | gravity acceleration |
| $\xi$ | rolling resistance coefficient (speed dependent) |
| $\alpha$ | slope angle |
| $J_p$ | powertrain moment of inertia |
| $\omega_k$ | wheel angular velocity |
| $v$ | vehicle speed |

Differential Eq. (3.13) is our basis for the vehicle energy consumption model. The last step is to calculate engine torque and power using gear ratio and powertrain efficiency (3.14) and finally to consider engine efficiency (3.15) (Fig. 3.15). With this model it is possible to calculate the energy consumption of the vehicle from inertial data and the actual vehicle state (speed, acceleration, slope angle, etc.) and with a small modification of the equation it is also possible to predict the speed of the car which was used in the coasting detection algorithm.

$$M_{mot} = \frac{\frac{M_h}{i_c}}{\eta_p} + M_{s\_mot} \tag{3.14}$$

$$P = \frac{M_{mot}}{\eta_{mot}} * \omega_{mot} \tag{3.15}$$

where

| | |
|---|---|
| $M_{mot}$ | engine torque |
| $M_h$ | driving torque |
| $M_{s\_mot}$ | engine inertia torque |

| $\omega_{mot}$ | engine angular velocity |
| $P$ | input power (energy consumption) |
| $i_c$ | gear ratio |
| $\eta_p$ | powertrain efficiency (gearbox, joints, etc.) |
| $\eta_{mot}$ | engine efficiency |

The model coefficients were identified on the same amount of measured data as for the optimal speed profile algorithm. The identification process was based on a deceleration test in three different modes:

(1) Free wheel mode—identification of rolling resistance and rotational inertia mass without engine (disconnected cutch).
(2) Free wheel mode with cornering—wheel resistance during cornering.
(3) Different shifted gears—engine inertia identification.

Example identification results are shown in Fig. 3.16. This simple model structure could be identified on-line in the vehicle during suitable maneuvers (coasting) for the adaptation to change some parameters during the life of the vehicle like increasing resistance in bearings etc.

### 3.2.4   Optimal Gearbox Shifting Map Design

In a real vehicle the automatic gearbox shifting is controlled by a shifting map. This map should be designed considering the vehicle dynamics and fuel/energy consumption. For a combustion engine it is possible to say when we need low fuel consumption the engine should make as low revolutions as possible. On the opposite side when we need a dynamic this engine should work at high revolutions because the power is increasing with the engine speed.

For an electric engine it is not so clear because of its behaviour. Consider the efficiency map for our asynchronous engine (see Fig. 3.15). This motor has the best efficiency between 3500 and 4500 rev/min and from 2200 rev/min has almost constant power—torque is decreasing with increasing revolutions. Due to this fact it is not possible to design a shifting map as same as for a combustion engine, but it is necessary to consider engine efficiency and power for various engine speeds and loads.

Consider the engine load only affected by a throttle position (driver command) and the engine speed as a function of the vehicle speed and selected gear ratio. With these conditions it is possible to calculate the shifting points to keep a higher engine efficiency or vehicle dynamics. Table 3.4 shows one example of shifting map where till 50% of throttle the shifting points are designed to keep higher efficiency and up to 50% to keep better vehicle dynamics. Numbers in Table 3.4 means the vehicle speed in kph. The green shifting points were designed for efficiency, blue points for dynamics, orange points were manually tuned for a smooth connection between efficiency and dynamics, red points are the speed limits of the map and black points are hysteresis for shift down.

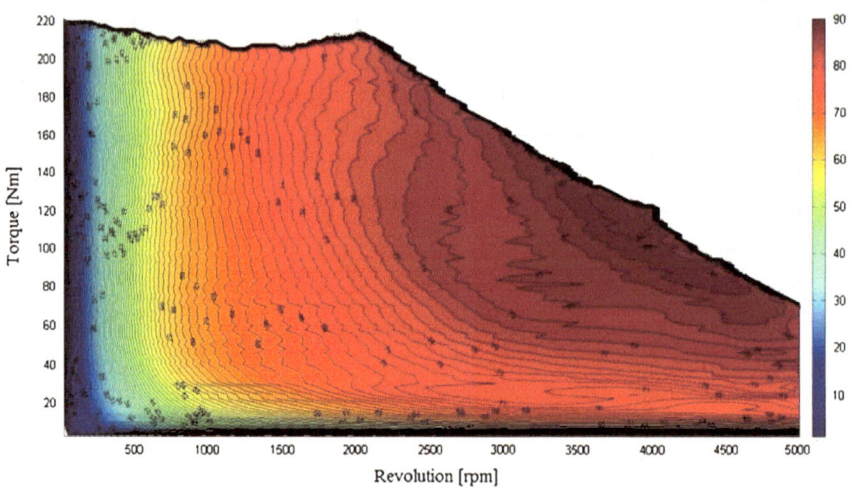

**Fig. 3.15** Powertrain efficiency map

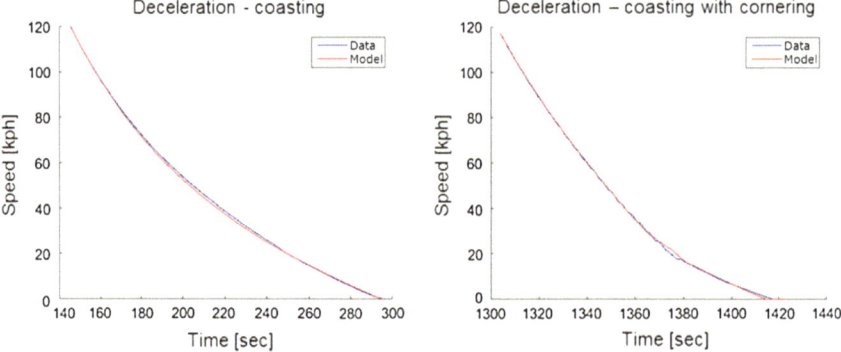

**Fig. 3.16** Example identification results

For a better description we show two shifting point calculations from 3rd to 4th (a) for throttle = 38% and (b) for 63%. For shifting point (a) the engine efficiency for 3rd gear is better up to vehicle speed 84 kph and from this speed it is a more effective 4th gear (Fig. 3.17). When the max engine revolution or torque limit is reached the efficiency and torque on wheel is shown with its last value. The torque on wheels (affects vehicle dynamics) is always higher for 3rd gear up to maximal revolution so this point is designed for efficiency.

The shifting point (b) it is better to design for dynamics considering the torque on wheels. The dynamics for the 3rd gear is better up to 82 kph and from this speed is better to use the 4th gear (Fig. 3.18).

**Table 3.4** Example shifting map for 7 speed automatic gearbox

| Shifting | Throttle position | | | | | | | | | | | | | | | | |
|---|---|---|---|---|---|---|---|---|---|---|---|---|---|---|---|---|---|
| | 0 | 6 | 13 | 19 | 25 | 31 | 38 | 44 | 50 | 56 | 63 | 69 | 75 | 81 | 88 | 94 | 100 |
| 1 -> 1 | 0 | 0 | 0 | 0 | 0 | 0 | 0 | 0 | 0 | 0 | 0 | 0 | 0 | 0 | 0 | 0 | 0 |
| 1 -> 2 | 20 | 20 | 23 | 23 | 24 | 27 | 28 | 29 | 35 | 35 | 34 | 33 | 31 | 30 | 30 | 30 | 30 |
| 2 -> 1 | 12 | 12 | 15 | 15 | 16 | 19 | 20 | 21 | 27 | 27 | 26 | 25 | 23 | 22 | 22 | 22 | 22 |
| 2 -> 3 | 36 | 36 | 38 | 45 | 45 | 59 | 59 | 58 | 58 | 58 | 55 | 53 | 51 | 49 | 49 | 49 | 49 |
| 3 -> 2 | 28 | 28 | 30 | 37 | 37 | 51 | 51 | 50 | 50 | 50 | 47 | 45 | 43 | 41 | 41 | 41 | 41 |
| 3 -> 4 | 49 | 49 | 50 | 62 | 64 | 68 | 84 | 87 | 92 | 88 | 82 | 79 | 77 | 77 | 77 | 77 | 77 |
| 4 -> 3 | 41 | 41 | 42 | 54 | 56 | 60 | 76 | 79 | 84 | 80 | 74 | 71 | 69 | 69 | 69 | 69 | 69 |
| 4 -> 5 | 84 | 84 | 87 | 91 | 93 | 128 | 131 | 135 | 128 | 122 | 114 | 110 | 106 | 106 | 106 | 106 | 106 |
| 5 -> 4 | 76 | 76 | 79 | 83 | 85 | 120 | 123 | 127 | 120 | 114 | 106 | 102 | 98 | 98 | 98 | 98 | 98 |
| 5 -> 6 | 112 | 112 | 112 | 113 | 117 | 126 | 157 | 160 | 164 | 154 | 147 | 144 | 135 | 135 | 135 | 135 | 135 |
| 6 -> 5 | 104 | 104 | 104 | 105 | 109 | 118 | 149 | 152 | 156 | 146 | 139 | 136 | 127 | 127 | 127 | 127 | 127 |
| 6 -> 7 | 140 | 141 | 143 | 145 | 147 | 148 | 180 | 213 | 202 | 190 | 180 | 172 | 165 | 165 | 165 | 165 | 165 |
| 7 -> 6 | 132 | 133 | 135 | 137 | 139 | 140 | 172 | 205 | 194 | 182 | 172 | 164 | 157 | 157 | 157 | 157 | 157 |
| 7 -> 7 | 255 | 255 | 255 | 255 | 255 | 255 | 255 | 255 | 255 | 255 | 255 | 255 | 255 | 255 | 255 | 255 | 255 |

Shifting point legend: efficiency, dynamics, manually adjusted for smooth connection of different designs, hysteresis, limits

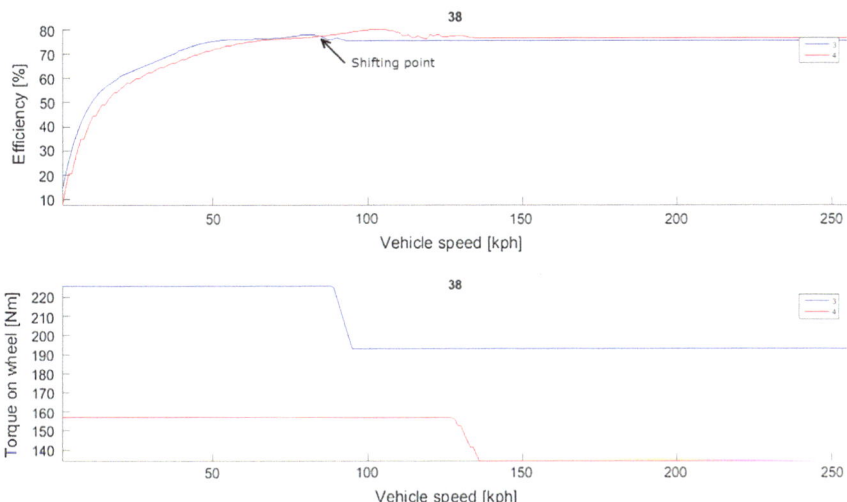

**Fig. 3.17** Shifting point from 3rd to 4th gear for throttle 38%

In this way it is possible to design the whole shifting map which is shown in Table 3.4. This shifting map was uploaded to a real gearbox and tested in defined cycles NEDC (Table 3.5) and WLTP (Table 3.6) in comparison with the constant gear realised with the same gearbox, but with a constantly shifted 4th gear. Tables 3.5 and 3.6 shows energy consumption for both cycles using shifting and without shifting.

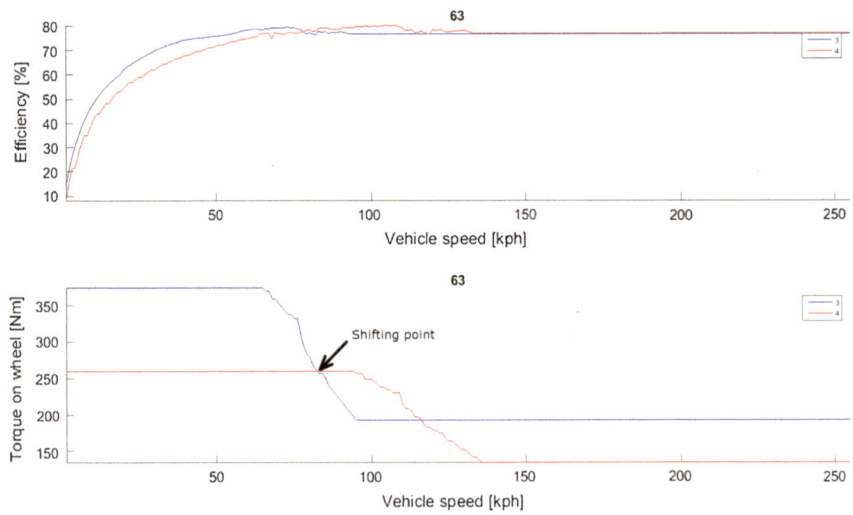

**Fig. 3.18** Shifting point from 3rd to 4th gear for throttle 63%

**Table 3.5** Results of NEDC cycle

| NEDC | | | | |
|---|---|---|---|---|
| Mode | Average speed [kph] | Energy pre cycle [kWh] | | |
| | | Consumed | Regenerated | Whole |
| Constant gear 4. | 34.62 | 1.96 | 0.14 | 1.82 |
| Shifting (1–6) | 34.33 | 1.81 | 0.15 | 1.66 |
| Benefit of gearbox [%] | | 7.43 | 12.89 | 8.94 |

**Table 3.6** Results of WLTP cycle

| WLTP | | | | |
|---|---|---|---|---|
| Mode | Average speed [km/h] | Energy pre cycle [kWh] | | |
| | | Consumed | Regenerated | Whole |
| Constant gear 4. | 46.81 | 4.28 | 0.32 | 3.96 |
| Shifting (1–6) | 46.88 | 4.14 | 0.37 | 3.78 |
| Benefit of gearbox [%] | | 3.16 | 15.29 | 4.63 |

## 3.3 Conclusion

During this work an algorithm for Optimal speed profile was designed and verified. This algorithm is independent to the type of vehicle, is optimal for a selected priority between energy consumption and average speed, considers safety limits and is applicable for advanced driver assistance systems and also for autonomous driving.

The demonstrated principle of coasting was also verified with a real car where the benefit of coasting for energy saving could be up to 12%, which of course it depends on the track.

The disadvantage of this principle is a very high sensitivity to track/map data quality because inputs to the Optimal speed profile design and model for speed predictions are track curvature, slopes and speed limits. In this case similar kinds of driver assistant systems should be connected with a well updated source of data and online information about traffic situations which could cause a speed limitation like road works.

For our goals, it was also necessary to design a new shifting map to consider an electric motor's behaviour because the Optimal speed design algorithm is not applicable on-line now and pre-designed shifting only cannot be used for real driving due to other unpredictable interruptions of driving like avoiding manoeuvers, breaking, etc. The benefit of shifting in comparison with a constantly selected gear in defined cycles NEDC (8.94%) and WLTP (4.63%) was also evaluated. The benefit for the real track near Mlada Boleslav was only simulated and it is approx. 1.3%. It is not so big a benefit because engine efficiency in 4th gear with an average speed over 80 kph is very high. Benefits also depend on driving cycles and the biggest benefit of shifting is expected in the cities where the FEVs make sense.

## References

1. Carmakers can free wheel to fuel efficiency targets, T&E report shows, 2013, Transport and enviroment. http://www.transportenvironment.org/press/carmakers-can-free-wheel-fuel-efficiency-targets-te-report-shows
2. Freewheel Function (Volkswagen). http://en.volkswagen.com/en/innovation-and-technology/technical-glossary/freilauffunktion.html

# Energy Efficient Driving in Dynamic Environment: Considering Other Traffic Participants and Overtaking Possibility

4

Zlatan Ajanović, Michael Stolz and Martin Horn

## 4.1 Introduction

Increasing environmental awareness, strict regulations on greenhouse gas emissions and constant desire to increase the range of electric vehicles as well as the big economic benefits drive a lot of research in the field of energy efficient driving. As a result, there are many different approaches addressing this topic. Some approaches are related to the vehicle design optimization, some to using alternative propulsion systems and some to the driving behavior optimization. In [1] the authors present a study which shows that the driving behavior has a rather big influence on energy consumption. It is shown that energy consumption may vary in a range of approx. 30% depending on moderate or aggressive driving behavior. Driving behavior related approaches for improving energy efficiency can be grouped into: "eco routing", "using road slope information", "traffic light assist", "platooning" and "overtaking" as shown in Fig. 4.1.

Eco routing is based on determining the most energy-efficient route for the trip, which may differ from the shortest or fastest one. E.g. [2] shows an example using historical data to determine an eco-route.

Z. Ajanović (✉) · M. Stolz
Area Electrics/Electronics and Software, Virtual Vehicle Research Center,
Inffeldgasse 21/A, 8010 Graz, Austria
e-mail: zlatan.ajanovic@v2c2.at

M. Stolz
e-mail: michael.stolz@v2c2.at

M. Horn
Institute of Automation and Control, Graz University of Technology,
Inffeldgasse 21/B, 8010 Graz, Austria
e-mail: martin.horn@tugraz.at

© The Author(s) 2017
D. Watzenig and B. Brandstätter (eds.), *Comprehensive Energy Management—Eco Routing & Velocity Profiles*, Automotive Engineering: Simulation and Validation Methods, DOI 10.1007/978-3-319-53165-6_4

**Fig. 4.1** Approaches to increase driving energy efficiency

Knowledge about the upcoming driving route, the road conditions and the ability to control the vehicle's propulsion enables the optimization of the speed trajectory of the vehicle with respect to the energy consumption. This problem has been extensively studied. Discrete dynamic programming (DP) for energy efficient driving has been used for over a decade now e.g. in research focused on heavy duty vehicles [3, 4]. A comparison between different optimization methods (Euler-Lagrange, Pontryagin's Maximum Principle, DP, and Direct Multiple Shooting) was presented in [5]. The reader will find there an analysis on the DP grid choice, tips on backward and forward dynamic programming, and on how to incorporate traffic lights. By using model predictive control (MPC) to drive vehicles on free roads with up and down slopes notable fuel savings are shown in [6]. MPC was also used to control a hybrid vehicle driving over a hill and performing vehicle following in [7]. In [8] an overview of the existing approaches and current state of the art can be found. Optimized speed trajectories are usually proposed to advice a human driver via an appropriate human-machine-interface (HMI). Rarely optimized speed trajectories are used to directly provide a reference value for underlying low-level controllers such as cruise control. Increasing vehicle automation is expected to change this in the near future.

The integration of time varying constraints such as traffic lights has also been studied intensively. Incomplete knowledge about upcoming traffic lights' timing was studied in [9], complete knowledge of the upcoming traffic lights' timing together with Dijkstra's algorithm was studied in [10] and an MPC based controller was developed with additional constraints imposed from a vehicle in front in [11]. MPC based controller proposed in [11] only considers vehicle following but not overtaking.

A lot of research on a topic of platooning was done within the SATRE project generating benefits for heavy duty vehicles [12].

In [13] a possible solution for a vehicle following problem is presented showing different concepts for safe vehicle following, defining helpful concepts such as the safe distance, time-inter-vehicular and time-to-collision. In [14] a possible solution for comfort oriented vehicle following with leading vehicle movement prediction treated as disturbance in MPC controller is presented. In publications dealing with the execution of optimal overtaking [15–18], speed trajectory planning is done in a way, that modifying an optimal speed trajectory leads to the smallest deviation from the desired speed while the vehicle is overtaking. These approaches are treating the problem locally and partially and don't give an energy consumption based decision if a vehicle should overtake or not and where the best location is for overtaking. Up to now not enough attention is paid on considering other participants in traffic especially leading vehicles.

If leading vehicles are neglected in the optimization, the unconstrained planning will not be fully achievable in real driving conditions, and may in some situations lead to drawbacks in energy consumption and very likely to bad driver acceptance. This work focuses on the integration of constraints imposed by leading vehicles and a global approach to optimize energy consumption.

## 4.2 Problem Definition

This chapter shortly states the problem definition from a mathematical point of view as an optimal control problem. The aim is to determine a vehicle speed trajectory (necessary motor torque) which results in optimal energy consumption for the given transportation task while taking into account additional constraints. First, a widely used, rather generic model of a vehicles longitudinal motion is defined. Then an appropriate cost function and relevant constraints are outlined. Finally, incorporation of time and/or space varying constraints (e.g. leading traffic) is discussed.

### 4.2.1 System Model

Since low model complexity is crucial for efficient optimization, the vehicle is modelled as particle-mass. This approach suits most applications and is the standard choice. The vehicle model (which we shortly call the system) is represented by two states: $s$—the longitudinal distance travelled by the vehicle and $v$—the longitudinal velocity of the vehicle. The system input is the motor torque $T_m$. The mathematical model is described by

$$\dot{s} = v, \tag{4.1}$$

$$\dot{v} = \frac{F_m(t)}{m} - \frac{F_r}{m}, \tag{4.2}$$

$$F_r = \frac{1}{2}\rho_a c_d A_f v(t)^2 + c_r mg\cos(\alpha(s(t))) + mg\sin(\alpha(s(t))). \tag{4.3}$$

The upper Eq. (4.1) originates from kinematics. Equation (4.2) is derived from Newton's second law with $F_m$ being the force produced by the propulsion motor, $m$ denoting the vehicle mass and $F_r$ being the resistive force. This resistive force is defined in (4.3), with air density $\rho_a$, aerodynamic drag coefficient $c_d$, the vehicle's frontal area $A_f$, rolling resistance coefficient $c_r$, gravity acceleration $g$ and the road slope angle $\alpha$. The propulsion element (which is here considered to be an electric motor with inner torque $T_m$) is modelled statically by:

$$F_m(t) = \frac{kT_m\eta^{sign(T_m(t))}}{r_w}, \tag{4.4}$$

using an efficiency coefficient $\eta$ scheduled by a map, a combined transmission ratio of the powertrain $k$ and the radius of the wheels $r_w$. The ratio between rotational speed $\omega$ of the motor and vehicle speed is defined by

$$\omega(t) = \frac{kv(t)}{2r_w\pi}. \tag{4.5}$$

## 4.2.2   Cost Function

Based on the system, in a next step a cost function is defined. In this contribution the main focus is on energy efficiency, so the cost function will be designed to represent the overall energy consumption of the vehicle.

If only energy used by the propulsion system of the vehicle is considered, the optimal solution generally speaking will most likely lead to rather slow movement. To avoid this, some authors additionally introduced a term to the cost function weighting travelling time [8, 13]. The weighting coefficient is then tuned such that the travel time is comparable to times achieved by human drivers. Although such an approach avoids slow movement, it results in a suboptimal solution from an energy efficiency point of view.

In the following, instead of introducing weighted travelling time we include power consumption from the auxiliary devices such as air conditioning system, infotainment system, thermal management system, etc. leading to straight forward energy related cost function. These auxiliary consumptions are assumed to be a constant load $P_{bn}$. The total power consumption of the vehicle is then the sum of boardnet power consumption and power consumption of the motor, which is calculated from the product of the rotational speed of the motor $\omega$ and motor torque

$T_m$. Consequently, the used energy is the integral of the power over time of the trip represented by

$$E_{\min} = \min_{T_m} \int_0^T \left( \omega(t) T_m(t) + P_{aux} \right) dt. \tag{4.6}$$

Instead of using time for integration, the distance can be used. This offers some distinct advantages for solving as final time is not known, but final distance is, and road slope appears as a function of distance [5].

### 4.2.3  Internal Constraints

Internal constraints in the optimization problem originate from constrained system dynamics, constraints on states, initial and final conditions. The following *internal* system constraints are considered:

Vehicle maximum speed and acceleration limits:

$$v_{\min} < v(t) < v_{\max}, \tag{4.7}$$

$$\dot{v}_{\min} < \dot{v}(t) < \dot{v}_{\max}, \tag{4.8}$$

Initial and final conditions for position and velocity:

$$v(0) = v_i, \quad v(T) = v_f, \tag{4.9}$$

$$s(0) = 0, \quad s(T) = S, \tag{4.10}$$

Note that the complete behavior of the electric motor such as maximum available torque, rotational speed and efficiency is modelled within the efficiency map. Apart from internal constraints, constraints which are imposed from environment and external factors exist.

### 4.2.4  External Constraints

External constraints can be classified according to the dependence on two variables relevant for the optimization problem. These are space and time. This means that external constraints can be grouped into four different groups summarized in Table 4.1.

**Table 4.1** External constraints classification

|                  | Time variant              | Time invariant                                  |
|------------------|---------------------------|-------------------------------------------------|
| Space variant    | Other traffic participants | Resting time (e.g. every 2 h)                   |
| Space invariant  | Traffic lights            | Traffic signs (e.g. speed limits), road curvature |

Generally speaking invariant constraints are easier to integrate into the optimization problem than variant constraints. In some cases time/space variant constraints can lead to tremendous efforts when considered.

*Time-and-space-invariant constraints*
This type of constraints is straight forward to integrate as it is constant in time and
space. Examples are: Traffic signs, which limit maximum speed on some road
segments or road curvature which also limits maximum speed because of the risk of
roll-over. On curved segments a minimum longitudinal acceleration limit can also
be imposed. The resulting acceleration would generate an inertial force pushing
back the driver and improving the drivability feeling since it may partly compensate
the uncomfortable centrifugal forces.

*Time-invariant-space-variant constraints*
An example of a constraint of this type could be resting time. Usually the driver has
to make resting stops after continuously driving for longer periods. This generally
doesn't explicitly depend on space as there are more resting spots along the road.
With some approximation, charging of the vehicle could be considered as a constraint of this type.

*Time-variant-space-invariant constraints*
Constraints of this type do not change in space but change with time. A typical
example is a traffic light, which sets the maximum speed to zero when the red light
is active. This happens in discrete time intervals at a fixed location in space.

*Time-and-space-variant constraints*
Constraints of this type are usually hard to consider, sometimes even impossible
with reasonable effort. Typical examples for a constraint of this type are other
vehicles moving in the surrounding traffic. Constraints are on the speed and position
of the controlled vehicle. Time and space of such constraints are not fixed, since the
velocity of the controlled vehicle itself influences the constraint. For example if the
controlled vehicle is moving faster, it will reach a leading vehicle sooner in time
and space. Things get even more complicated when considering the possibility of
overtaking. In this case the speed of the controlled vehicle has to be significantly
higher than the speed of the leading vehicle, providing a speed difference to safely
overtake. Such constraints can be mathematically expressed as:

$$|s(t) - s_{lead}(t)| > d_{safe} \qquad (4.11)$$

$$v(t) - v_{lead}(t) > v_{safe}. \qquad (4.12)$$

Constraint (4.11) is active in case of vehicle following and constraint (4.12) in
case of overtaking a leading vehicle.

From a practical point of view, as the future movement of other vehicles is
usually unknown, some assumptions must be made. In this work, the most simple

movement prediction is used by assuming, that other vehicles will continue to move with their actual speed. Additionally, it will be assumed that other vehicles will not overtake the controlled vehicle and may slow down behind in order to avoid collisions once they have been overtaken. This assumption leads to the possibility of neglecting vehicles following the controlled vehicle.

## 4.2.5  Challenges

There are a few challenges connected to the discussed optimization, which are shortly highlighted in the following:

*Nonlinear problem*
The problem is nonlinear, as air drag resistance is proportional to the square of the velocity. Including the motor efficiency adds additional nonlinearity, as there is a sign function in motor torque calculation.

*Dynamic constraints*
The incorporation of the dynamic constraints increases the problem complexity. Many methods such as backward dynamic programming with free final time cannot cover them as these constraints depend on the trajectory from the start.

*Course of dimensionality*
If the optimization is solved numerically, state discretization is necessary. Increasing the number of system states considered in optimization problem dramatically increases the number of possible state combinations. Each additional state multiplies the number of state combinations by the number of its discretization values. E.g. Considering multiple propulsion sources as in hybrid cars, or even considering distinctive gear ratios will highly increase the computational effort for solving.

*Real-time computation versus precision*
As the focus is on real-time applications within a vehicle the general tradeoff between speed of computation and precision has to be taken into account. This will have an impact on the choice of methods.

## 4.3  Optimal Motion Planner

As explained above, publications considering road slope to generate energy efficient speed trajectory do not fully consider other vehicles in traffic and the possibility to overtake them. On the other hand publications dealing with optimal overtaking tackle the problem locally and consider efficiency only as deviation from a desired speed which leads to a local optimal solution.

In contrary to existing approaches, this work aims to achieve a global optimal solution by generating a new optimal speed trajectory. A new optimal speed trajectory has to be generated since the constraints are assumed to be violated. Therefore, the unconstrained optimal speed trajectory in this case is not optimal and there is no argument in persisting to still track it.

The approach proposed in this work gives valid conclusion if it is more beneficial (in terms of overall energy consumption) to overtake a leading vehicle or to slow down based on predicted energy usage for the trip and actual overtaking maneuver and on which segment of the road to overtake. The planning is based on a motion prediction (assumption) of the leading vehicle moving in front and information about the upcoming road.

In this work dynamic programming is used as a method to derive the optimal control (here the speed trajectory) because of its broad range of applicability and the flexibility to integrate different constraints.

### 4.3.1  Dynamic Programming

Dynamic programming is a very common method used for solving the optimal control problem discussed in this work. The main advantages are its flexibility and possibility to incorporate different kinds of models and constraints. It is based on the Principle of Optimality, and it was introduced by Bellman [19]. The interested reader is referred to [20].

*Principle of Optimality*
*An optimal policy has the property that whatever the initial state and initial decisions are, the remaining decisions must constitute an optimal policy with regard to the state resulting from the first decisions* [19].

This principle enables iterative search for optimal solution starting from the end point and building an optimal trajectory to the start. In each step transitions from all possible current states to the all previous states with accumulated previous values are calculated and minimum costs and respective transitions are selected for each possible current state. In this way going backwards in time, trajectories are growing by using calculation results from the "previous" step.

For this work a tailored and computationally optimized solution for optimal speed trajectory planning based on dynamic programming was developed in MATLAB making intensive use of matrix calculus. For a trip of **4200 m** with discretization steps $ds = 5$ **m** and $dv = 0.1$ **m/s** solving on a standard PC (intel i5) it takes **8.83 s** to plan an optimal speed trajectory using energy efficiency maps for the electric motor and **4.81 s** using a constant efficiency.

To validate the implementation both, forward and backward dynamic programming schemes were implemented. The achieved results are identical, as it was expected. The advantage of the backward calculation is that the calculated result can be reused during the trip as it only depends on the final state. This is not the case with the forward calculation, where results are related to the specific initial state. On the other hand the advantage of the forward calculation is that other states such as the position of other vehicles can be calculated as the initial time is always known.

The implementation of dynamic programming can use time or distance as a basis for discretization. Using distance is useful in finding the energy optimal trajectory as the final time is not fixed but it has several disadvantages such as, when the speed is zero it is impossible to calculate time. An additional disadvantage is that on high speeds time shortens and with fixed speed discretization steps the number of possible transitions which satisfy the maximum acceleration constraints decreases. Nevertheless, an appropriate selection of the discretization steps leads to a valid solution. Therefore, distance as a base is used in this implementation.

In order to deeply understand the optimal control discussed in this work we introduce two helpful concepts: the optimal trajectory tree and the cost-to-go map.

### 4.3.2   Optimal Speed Trajectory Tree

We define the so called optimal speed trajectory tree as a tree like structure formed by connecting all optimal transitions by a line. This results in a nice representation of the optimal control problem and together with a cost-to-go map it gives insight into an optimal behavior of the case without leading vehicles. This will be useful in understanding the advantage of the proposed approach compared to existing approaches, which are discussed in Sect. 4.4.2.

What can be noted is that generally if two different trajectories have a common node they will continue on same trajectory towards the goal.

An optimal speed trajectory tree for a problem considered in this work with discretization steps of 10 m for distance and 0.3 m/s for speed is shown on Fig. 4.2. This map is generated from end to front in backwards DP. Additionally to the optimal solution for the given initial condition, we can see optimal solutions for different initial conditions and the impact of different initial conditions.

### 4.3.3   Cost-to-Go Map

The so called cost-to-go map represents the energy needed to finish a trip on an optimal trajectory from each point in distance-velocity plane. In Fig. 4.3 the cost-to-go map for the same problem as in Fig. 4.2 is shown for the backward DP approach.

The characteristics of the cost to go map are closely related to the cost function and it can be noted that costs increase as distance is closer to the start. Additionally, costs are smaller on higher speeds because of the bigger kinetic energy stored in a moving vehicle. Costs are also smaller on a hill and bigger in the valley because of the potential energy.

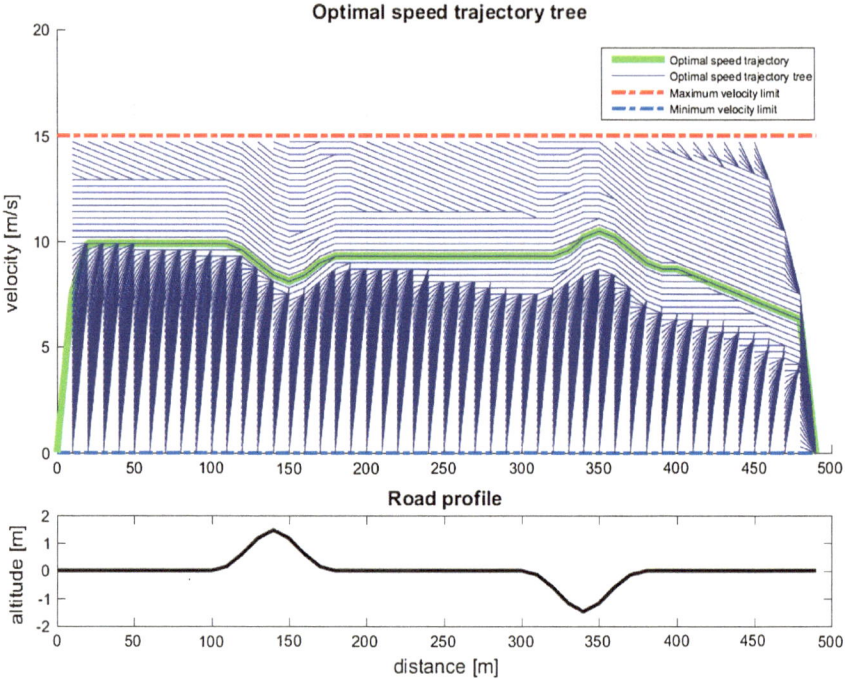

**Fig. 4.2** Optimal speed trajectory tree in backward planning dynamic programming

### 4.3.4 Considering External Constraints

*Time-and-space-invariant constraints*
As mentioned in Sect. 4.2 examples of constraints of this type are: traffic signs, which limit the maximum speed on some road segment or due to road curvature which also limits the maximum speed because of the risk of roll-over. These constraints can be easily incorporated into the optimization and can be handled by in both, forward and backward DP, if distance is used as the integration variable. This type of constraint imposes that transitions to the speed values which are conflicting with a traffic sign are not allowed on the distance segments where the traffic sign is active. Impossible transitions are implemented by assigning an infinite cost value to that transition.

*Time-and-space-variant constraints*
Constraints of this type are a main focus of this work, since it deals with other vehicles moving in the leading traffic. These constraints are imposed on speed and position of the controlled vehicle. As mentioned previously, both time and space of such constraints are not fixed, because the velocity of the controlled vehicle itself influences this constraint.

Due to this, forward DP is used here, enabling to calculate the distance between the controlled vehicle and the leading vehicle for each trajectory, which is possible

**Cost-to-go map: Energy [kJ]**

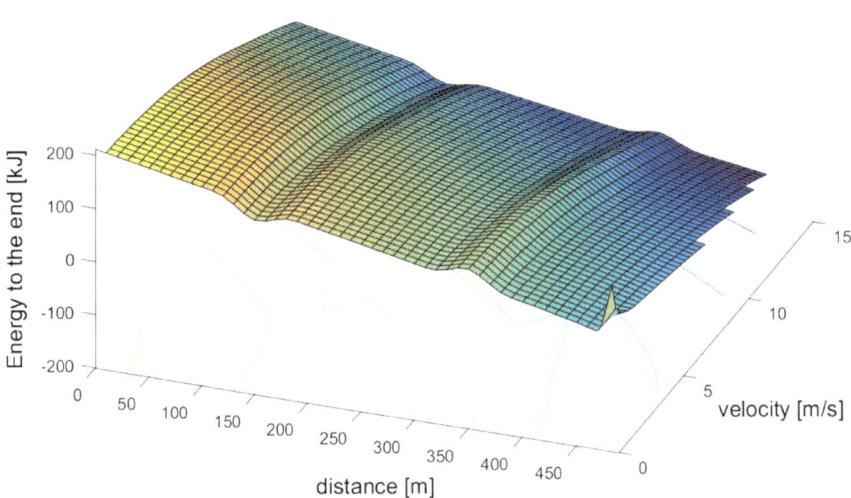

**Fig. 4.3** Cost-to-go map of cumulative costs to travel to the end (backward planning)

since the initial distance is known. The movement of the other vehicle is calculated using a simple prediction model that assumes that the leading vehicle is moving with constant speed and that it will slow down if it reaches the controlled vehicle (after being overtaken). Note that a more sophisticated model of the leading vehicle's choice of velocity which may depend on space, time and the controlled vehicle can be easily included.

Finally, the transitions which do not satisfy the constraints (4.11) or (4.12) explained in Sect. 4.2 are defined to be impossible. Disabling a transition will disable all trajectories leading to a collision. As a consequence only trajectories which do not lead to a collision will be checked and the best out of these will be selected.

*Time-variant-space-invariant   constraints   and   time-invariant-space-variant constraints*

Constraints of this type are not considered now, they will be tackled in future works. These constraints can be considered as a special case of time-and-space-variant constraints, therefore incorporating them should be straight forward once time-and-space-variant constraints are incorporated.

## 4.4   Simulation Results

In this section, analyses and comparison to other approaches will be presented in order to highlight fundamental attributes and advantages of the proposed approach. Discussion of the results will be based on the vehicle trajectories in distance-velocity and time-distance plots, appropriate tables will show the differences in consumed energy as well as travelled time.

In the first subsection, driving with constant speed is compared to optimal driving. In the second subsection, differences between the proposed and existing overtaking approach will be investigated. In the third subsection, a short analysis of the influence of the road gradient on the overtaking event will be made. After that, the influence of the speed difference on overtaking event will be analyzed.

For this purpose, a road segment of 500 m with a hill and a valley will be used. The hill is 1.6 m high with maximum slope of 6%; the valley has the same shape but negative. Discretization step size for distance is 1 m and for speed 0.03 m/s have been used. If not stated differently, this discretization applies to all simulation experiments.

### 4.4.1   Constant Speed Versus Optimized Variable Speed

At first driving with constant speed is compared to driving at optimal variable speed. The example simulation is setup as follows: Three different speed set points (7, 9, 11 m/s) are used. The vehicle starts from standstill and accelerates to reach the set speed. Then it continues to travel with this speed until it has to decelerate to finally stop at the end of the trip.

In contrast to this, optimal driving, as described in Sects. 4.2 and 4.3, takes road slope, air resistance, roll coefficient, board net power consumption, and powertrain efficiency into consideration. An (in the sense of energy usage) optimal speed trajectory is generated for the given travel.

As can be seen in Fig. 4.4, this results in moving on a constant speed of round about 10 m/s (for this vehicle and air drag resistance) on a flat road, slowing down uphill and speeding up downhill.

Table 4.2 summarizes the results from Fig. 4.4 with respect to energy consumption and traveling time. Concluding this example, an improvement in energy consumption above 8% can be observed. Prolongation of the trip is only about 1.3%.

### 4.4.2   Proposed Versus Existing Overtaking Approach

In this subsection, the proposed approach is compared to existing overtaking approach. As mentioned in the introduction in literature optimal overtaking is usually incorporated as a least quadratic (LQ) deviation from the desired vehicle

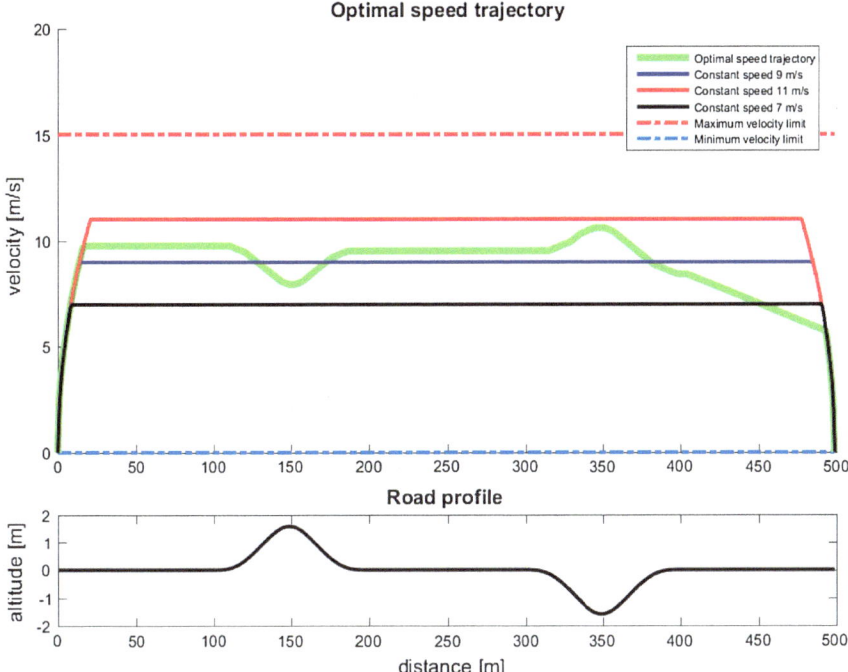

**Fig. 4.4** Constant speed versus optimized variable speed

trajectory. This means that an unconstrained optimal speed trajectory is generated and then modified to satisfy safety requirements and avoid collisions.

In the following simulation example, it is assumed that a leading vehicle will move with a constant speed of 8 m/s located initially 20 m ahead of the controlled vehicle. In Fig. 4.5 the green speed trajectory represents the optimal speed trajectory for the unconstrained case. This trajectory is used as reference trajectory for generating a collision free trajectory (red color) as proposed in literature. In contrast to this, the blue trajectory is the optimal speed trajectory resulting from directly incorporating the constraint into the optimization problem.

As expected from Fig. 4.5, Table 4.3 reveals that regarding the energy consumption as well as traveling time considering the constraint in the optimization leads to better results. The constrained optimization leads to only slightly larger energy consumption (+0.74%) compared to the unconstrained case. The standard approach leads to a bigger difference (+5.27%). This implies that integration of the leading vehicle as a constraint into the optimization problem may play a considerable role in reducing energy consumption. Even there is the potential to additionally save traveling time.

Note that Fig. 4.6 shows that the location and point in time of the overtaking event for both approaches differ. In the case of constrained planning, the overtaking

**Table 4.2** Constant speed versus optimized variable speed

|                | Energy used [kJ] | Difference [%] | Time travelled [s] | Time difference [%] |
|----------------|------------------|----------------|--------------------|---------------------|
| Optimal traj.  | 213.87           | 0              | 59.25              | 0                   |
| Speed 9 m/s    | 231.68           | +8.3           | 58.49              | −1.28               |
| Speed 11 m/s   | 237.98           | +11.3          | 49.06              | −17.20              |
| Speed 7 m/s    | 237.97           | +11.3          | 73.75              | +24.47              |

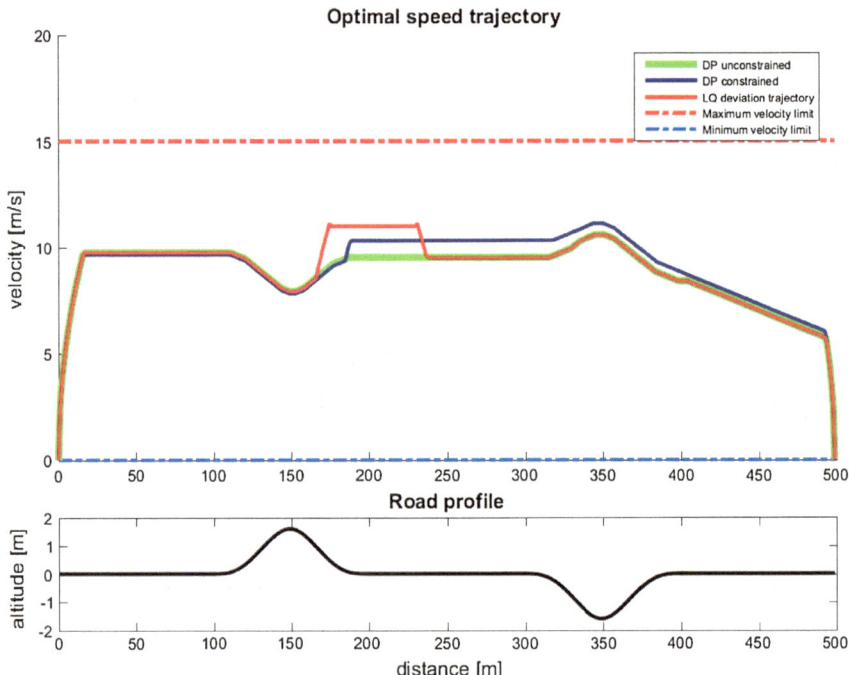

**Fig. 4.5** Proposed versus existing overtaking approach

**Table 4.3** Proposed versus existing overtaking approach

|               | Energy used [kJ] | Difference [%] | Time travelled [s] | Difference [%] |
|---------------|------------------|----------------|--------------------|----------------|
| Unconstrained | 213.87           | 0              | 59.25              | 0              |
| Constrained   | 215.45           | +0.74          | 57.34              | −3.23          |
| LQ deviation  | 225.15           | +5.27          | 58.27              | −1.66          |

Summary on energy consumption and traveling time

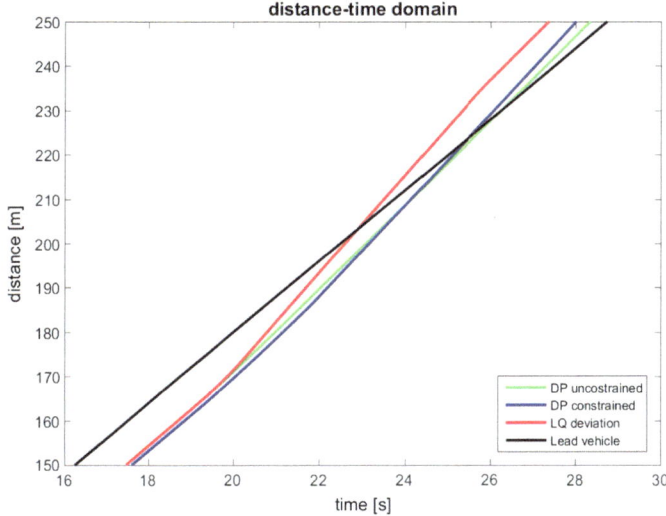

**Fig. 4.6** Different locations of overtaking events in proposed and existing overtaking approach

event is at 224 m. In the case of LQ deviation planning overtaking event is at 202 m.

Computation time for calculating LQ deviation trajectory can be approximated to be almost twice as big as necessary, because first the unconstrained planning has to be generated and then LQ deviation trajectory has to be calculated.

### 4.4.3 Influence of the Road Gradient

Within another simulation study the influence of the road gradient on a constrained trajectory will be analyzed. The leading vehicle's initial position will be varied so that the leading vehicle trajectory crosses the unconstrained trajectory at different locations (possible collision at different locations). The leading vehicle will move with a constant speed of 8 m/s and its initial position will be set to 5, 10, 15, and 20 m.

As shown in Fig. 4.7 the algorithm doesn't give uniform results for all cases and it seems that there are some preferred segments for overtaking. Either the vehicle speeds up and overtakes the leading vehicle before the hill or it slows down and waits until it passed the hill and overtakes on a downhill section (red trajectory). This complies with usual human driving behavior (at least when considering energy consumption).

Table 4.4 shows that energy consumption in all of the cases is not much bigger than 1% and time is smaller as a result of speeding up, compared to the unconstrained problem.

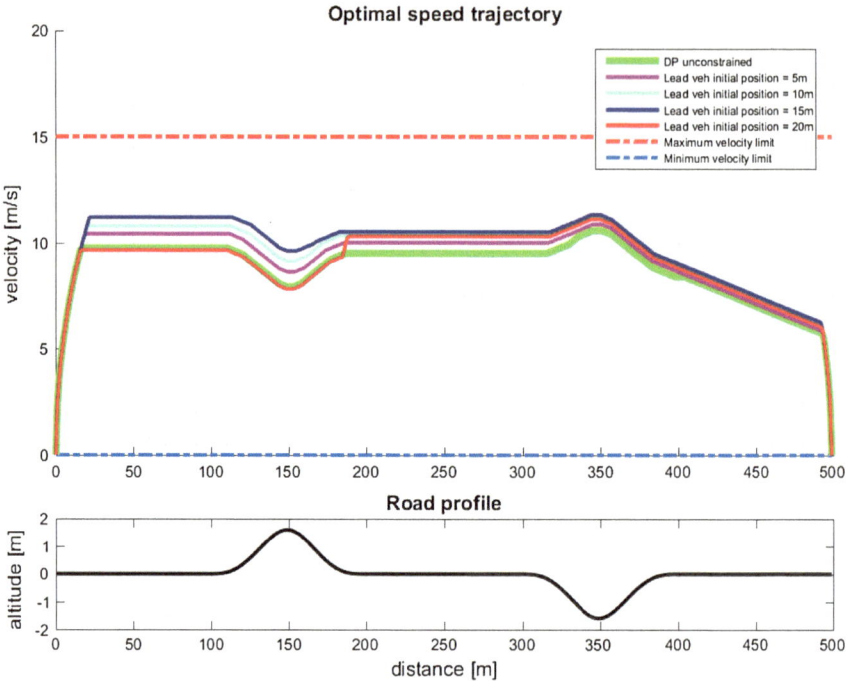

**Fig. 4.7** Influence of the road gradient

### 4.4.4 Influence of the Leading Vehicle Speed

In a last simulation study the influence of the leading vehicle speed on a constrained trajectory will be analyzed. For this reason, the leading vehicle's speed will be varied. The leading vehicle's position is also adjusted so that the leading vehicle's trajectory crosses the unconstrained trajectory always at 250 m. The leading vehicle will have constant speeds with 7.5, 8, 8.5, and 9 m/s.

Figure 4.8 shows the simulation results of the above described setup. It can be seen that the optimization again doesn't give uniform results for all cases and that sometimes it is better to slow down and follow the leading vehicle (blue and red) and sometimes to speed up and overtake (magenta and cyan). This also complies with usual human driving behavior. Table 4.5 shows that differences in energy consumption in all of the cases are rather small and differences in travel time depend on following or overtaking.

**Table 4.4** Influence of the road gradient

|  | Energy used [kJ] | Difference [%] | Time travelled [s] | Difference [%] |
|---|---|---|---|---|
| Unconstrained | 213.87 | 0 | 59.25 | 0 |
| Init. pos. = 0 m | 214.26 | 0.18 | 56.78 | −4.17 |
| Init. pos. = 10 m | 214.99 | 0.52 | 55.12 | −6.98 |
| Init. pos. = 15 m | 216.07 | 1.03 | 53.85 | −9.12 |
| Init. pos. = 20 m | 215.45 | 0.74 | 57.34 | −3.23 |

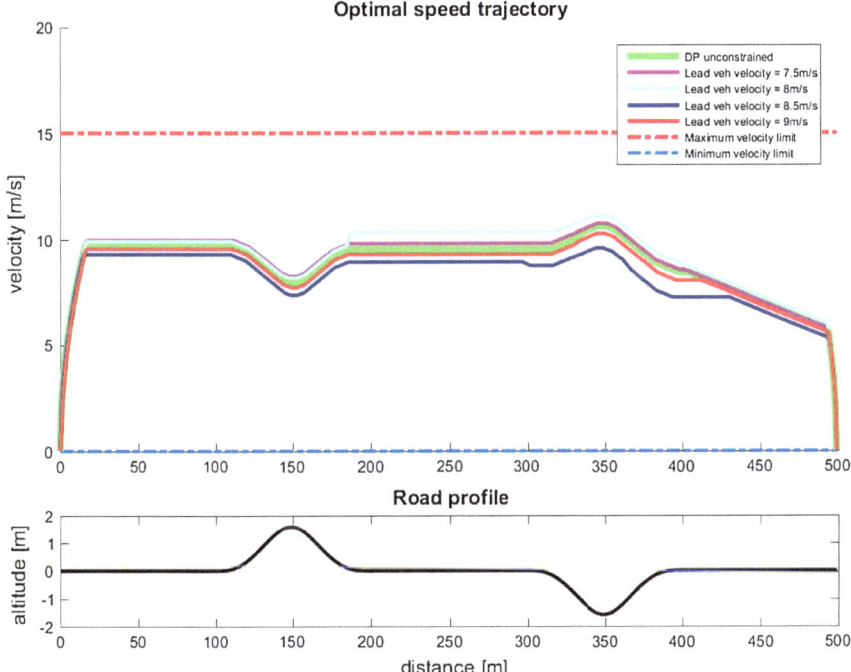

**Fig. 4.8** Influence of the leading vehicle speed

**Table 4.5** Influence of the leading vehicle speed

|                 | Energy used [kJ] | Difference [%] | Time travelled [s] | Difference [%] |
| --------------- | ---------------- | -------------- | ------------------ | -------------- |
| Unconstrained   | 213.87           | 0              | 59.25              | 0              |
| LV velocity 7.5 | 213.94           | 0.03           | 58.07              | −1.99          |
| LV velocity 8   | 215.02           | 0.53           | 56.80              | −4.13          |
| LV velocity 8.5 | 215.61           | 0.81           | 63.49              | 7.16           |
| LV velocity 9   | 215.42           | 0.72           | 61.56              | 3.89           |

## 4.5 Conclusion

Forward dynamic programming, as a flexible approach, allows the consideration of available movement prediction of leading vehicles as constraints within travel speed optimization. As a result, a global optimal solution in energy consumption can be obtained. Sometimes additional travel time reduction compared to the existing overtaking approach reported in literature has been observed.

The constrained problem showed up to require much smaller discretization steps to produce valid results compared to the unconstrained problem. A reason for this is that the optimal trajectory is usually very close to the constraints during an overtaking maneuver so a coarse grid can cause unwanted oscillations. As the minimum clearance between vehicles is only a few meters, it is also important that the space discretization step is at least one order smaller. Small space discretization steps automatically imply small velocity discretization steps to provide enough possible transitions in each step.

For the sake of simplicity, the presented work investigated one leading vehicle moving in the same direction on a two lane road. However, extending the optimization to handle more vehicles or vehicles moving in different directions is straight forward, since additional vehicles are represented by constraints of the same type. The additional constraints just increase the computational effort, but the calculation schema stays the same. The consideration of available lanes at certain segments can be accomplished by adding the lane as an additional state.

Although considering other traffic participants within the optimization increases the effort for solving compared to the original unconstrained problem, there is a big potential for improvements in both, energy consumption and travelling time.

First simulation studies also show that driver acceptance will not be a problem since the behavior is intuitive.

**Acknowledgements** The project leading to this study has received funding from the European Union's Horizon 2020 research and innovation programme under the Marie Skłodowska-Curie grant agreement No 675999, ITEAM project.
VIRTUAL VEHICLE Research Center is funded within the COMET—Competence Centers for Excellent Technologies—programme by the Austrian Federal Ministry for Transport, Innovation and Technology (BMVIT), the Federal Ministry of Science, Research and Economy (BMWFW),

the Austrian Research Promotion Agency (FFG), the province of Styria and the Styrian Business Promotion Agency (SFG). The COMET programme is administrated by FFG.

# References

1. Bingham C, Walsh C, Carroll S (2012) Impact of driving characteristics on electric vehicle energy consumption and range. IET Intel Trans Syst 6(1):29–35
2. Minett CF, Salomons AM, Daamen W, Van Arem B, Kuijpers S (2011) Eco-routing: comparing the fuel consumption of different routes between an origin and destination using field test speed profiles and synthetic speed profiles. In 2011 IEEE forum on integrated and sustainable transportation system (FISTS), Vienna
3. Hellström E (2005) Explicit use of road topography for model predictive cruise control in heavy trucks. MS thesis, Linkoping University, Sweden
4. Hellström E (2010) Look-ahead control of heavy vehicles. Ph.D. thesis, Linköping, Sweden
5. Saerens B (2012) Optimal control based eco-driving. Ph.D. thesis, Katholieke Universiteit Leuven, Leuven
6. Kamal MAS, Mukai M, Murata J, Kawabe T (2011) Ecological vehicle control on roads with up-down slopes. IEEE Trans Intell Transp Syst 12(3):783–794
7. Vajedi M, Azad NL (2016) Ecological adaptive cruise controller for plug-in hybrid electric vehicles using nonlinear model predictive control. IEEE Trans Intell Transp Syst 17(1):113–122
8. Sciarretta A, Nunzio GD, Ojeda L (2015) Optimal ecodriving control: energy-efficient driving of road vehicles as an optimal control problem. IEEE Control Syst Mag 71–90
9. Mahler G, Vahidi A (2012) Reducing idling at red lights based on probabilistic prediction of traffic signal timings. In: 2012 American control conference (ACC), Montreal
10. De Nunzio G, Wit CC, Moulin P, Di Domenico D (2013) Eco-driving in urban traffic networks using traffic signal information. In: 52nd IEEE conference on decision and control, Florence, Italy, 10–13 Dec 2013
11. Kural E, Jones S, Parrilla AF, Grauers A (2014) Traffic light assistant system for optimized energy consumption in an electric vehicle. In: International Conference on Connected Vehicles and Expo (ICCVE)
12. The SARTRE Project. [Online]. Available: http://www.sartre-project.eu. Accessed 11 Aug 2016
13. Mensing F, Bideaux E, Trigu R, Tattegrain H (2013) Trajectory optimization for eco-driving taking into account traffic constraints. Trans Res Part D Trans Environ 18:55–61
14. Schmied R, Waschl H, del Re L (2016) Comfort oriented robust adaptive cruise control in multi-lane traffic conditions. In: 8th IFAC international symposium on advances in automotive control, Norrköping, Sweden, 2016
15. Wang M, Hoogendoorn S, Daamen W, van Arem B, Happee R (2015) Game theoretic approach for predictive lane-changing and car-following control. Transp Res Part C Emerg Technol 58(Part A):73–92
16. Murgovski JSN (2015) Predictive cruise control with autonomous overtaking. In: 54th IEEE conference on decision and control (CDC), Osaka, Dec 2015
17. Kamal MAS, Taguchi S, Yoshimura T (2016) Efficient vehicle driving on multilane roads using model predictive control under a connected vehicle environment. IEEE Trans Intell Transp Syst (99):1–11
18. Shamir T (2004) How should an autonomous vehicle overtake a slower moving vehicle: design and analysis of an optimal trajectory. IEEE Trans Autom Control 49:607–610

19. Bellman R (1954) The theory of dynamic programming. The Rand Corporation, Santa Monica
20. Bertsekas D (2007) Dynamic programming and optimal control. Athena Scientific

# Model-Based Eco-Routing Strategy for Electric Vehicles in Large Urban Networks

**5**

Giovanni De Nunzio, Laurent Thibault and Antonio Sciarretta

## 5.1 Introduction

While research on energy-efficient driving trajectories (i.e. eco-driving) has already witnessed numerous efforts [1], the impact analysis of route choice on energy consumption and the development of energy-efficient navigation systems (i.e. eco-routing) are relatively new subjects. Current available navigation systems suggest either the fastest or the shortest route. However, these route choices are not always the most energy-efficient ones [2]. A recent experimental study in Sweden found that the drivers' spontaneous route choice is not the most fuel-efficient for 46% of trips in the city of Lund. Also, these trips could have saved 8% fuel by using a fuel-optimized navigation system [3].

The main challenge of an eco-routing navigation system lies in the accurate estimation of the vehicle energy consumption, as demonstrated in [4, 5]. Such energy consumption should be estimated on each of the different road segments composing the routing network of the area under analysis. Models for estimating vehicle energy consumption can be broadly divided into macroscopic and microscopic models, depending on how vehicular activities are aggregated over time and space. In a comprehensive study in [5], a comparison of several existing energy consumption and emission models for combustion-engine vehicles was conducted. After comparing 11 state-of-the-art consumption models, the authors conclude that

G. De Nunzio (✉) · L. Thibault · A. Sciarretta
Department of Control, Signal and System, IFPen, Rueil-Malmaison, France
e-mail: giovanni.de-nunzio@ifpen.fr

L. Thibault
e-mail: laurent.thibault@ifpen.fr

A. Sciarretta
e-mail: antonio.sciarretta@ifpen.fr

© The Author(s) 2017
D. Watzenig and B. Brandstätter (eds.), *Comprehensive Energy Management—Eco Routing & Velocity Profiles*, Automotive Engineering: Simulation and Validation Methods, DOI 10.1007/978-3-319-53165-6_5

81

only few may be effectively used for assigning eco-weights, among which also some macroscopic models.

In general microscopic models are considered more precise because they take as input instantaneous velocities and accelerations, which are difficult to obtain without equipped probe vehicles recording GPS and/or CAN-bus data. Eco-routing navigation systems based on this type of models need to acquire a large amount of driving data in order to determine a statistical cost on each road segment of the routing model. For instance, in [6] the authors propose a strategy to reconstruct synthetic speed profiles from historical speed data to calculate an instantaneous power demand and an energy cost for the road segments. In [7], analogously, the authors propose to compute eco-weights based on GPS data, and to periodically maintain these weights as new information is collected. Intuitively, this approach requires a long time to have a reliable and dense mapping of the energy costs of all the road segments in the routing network. Also, it requires arbitrary decisions to deal with data sparsity and incompleteness, and it is not suited for real-time route planning because it is limited to the GPS data area coverage.

On the other hand, macroscopic models take as input mean velocities, mean travel time and road grade, which are typically easier to obtain through free or commercial historical databases. Such aggregated models can be further classified into regression-based models and load-based models. The regression-based models are generally difficult to calibrate in order for all the parameters of the model to be significant, and they may lead to random errors which are not easy to explain due to the little physical foundation of these models [8–10]. The load-based models rely upon the longitudinal dynamics of the vehicle and are easier to calibrate using the vehicle construction parameters [11, 12].

All these models make strong assumptions for the energy consumption estimation, by neglecting accelerations, or road grade, or the impact of the road network topology and signalization. Only few recent works address the importance of considering accelerations even when using macroscopic models, as well as considering the elements of the road infrastructure that may cause vehicular speed disruptions. In [13], the authors propose a macroscopic load-based model to compute the eco-weights on the road segments. They also propose a method to consider inter-link accelerations induced by the different adjacent average speeds. To do so, they have the intuition to separate the costs for all the turning movements available at a certain intersection. However, though aware of the fact that the size of the routing problem could grow significantly, they neglect the issue. In addition, since electric vehicles and energy recovery phenomena are considered, the eco-routing strategy has to deal with the additional challenge of negative energy weights on certain road segments of the routing network. Optimal path-searching algorithms on graphs with negative weights (e.g. Bellman-Ford algorithm [14]) are often discarded in literature because of their computational cost. Alternative approaches have been proposed to tackle this issue, mostly based on heuristics for graph weights shifting [12] or non-optimal path searching [15].

In this work, a novel macroscopic energy consumption model and a novel eco-routing strategy are described. The novelty of the proposed approach is summarized as follows.

First, the macroscopic energy consumption model, based on the real physics of the vehicle, considers accelerations to move from one road segment to the adjacent ones. Also, additional accelerations induced by the road topology and infrastructure (e.g. traffic lights, stop signs, turns, etc.) are considered in the energy consumption model.

Then, since considering accelerations introduces a correlation between the energy cost of adjacent links, and the cost may not be unique due to the presence of multiple turning movements, this ambiguity is solved by deploying the routing strategy on the weighted *adjoint* graph. Such graph allows to decouple all the turning movements for a correct and non-ambiguous energy cost assignment, and also the decoupling is efficiently achieved avoiding the exponential growth of the routing graph.

Finally, the presence of negative energy weights, due to regenerative breaking, is addressed by using an optimal path-searching algorithm, namely Bellman-Ford algorithm. The computational time is shown to remain acceptable, and also a strategy for a real-time use of the eco-routing navigation system is detailed.

The overall architecture is arranged as a three-layer structure, where the (1) acquisition of road network and traffic data, the (2) vehicle energy consumption model, and (3) the energy-oriented navigation algorithm layers all contain improvements with respect to the state of the art. Finally, validation results obtained via an experimental campaign are presented.

## 5.2 Problem Description

The objective of this work is a reduction of the energy spent for traction by electric vehicles, therefore an increase of their range, obtained through an intelligent navigation system that performs an energy consumption minimization given an origin and a destination. This may be seen as an optimization problem where the function to be minimized is the energy consumption of the vehicle. The objective function depends on several variables, different in nature and difficult to know or estimate: vehicle parameters, network topology, traffic conditions, road grade, driving style, etc.

In Fig. 5.1, the proposed eco-routing functioning scheme is reported. Mainly, a reliable and effective eco-routing strategy depends on three critical modules.

The quality of the *map data* is of primary importance for a realistic representation of the road transportation network, on which the navigation strategy will be deployed. These data can be static and dynamic. Static map data include position of the roads, position of the intersections, driving directions on the roads, road grade. Dynamic map data include traffic conditions. An accurate *energy consumption model* allows to effectively predict the energy consumption over a given

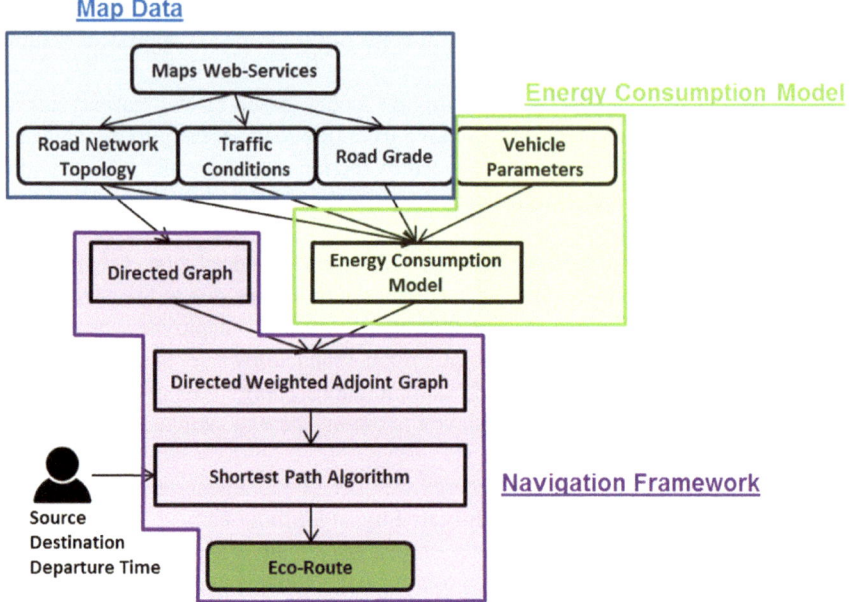

**Fig. 5.1** Overview of the proposed eco-routing strategy

trip. Finally, the map data can be used to generate a model of the road transportation network and the *navigation framework*. Such model is fed with the energy consumption estimates given by the vehicle model in order to assign a cost to each road section in the model. An algorithm to find the energy-optimal sequence of road segments to drive from the desired origin to the desired destination is needed to return the eco-route.

In the following, each module of the strategy will be detailed.

## 5.2.1 Map Data

The accuracy and reliability of the macroscopic information about the road transportation network and traffic conditions are paramount for an effective energy consumption prediction and a credible energy-aware navigation assistance.

For this work, the data were provided by Here Maps web-services [16].

### Road Network Topology
The position of road segments and intersections, the driving directions on the road segments, and the altitude along the road segments are what is usually referred to as "road topology". The information about the road network topology is provided by Here Maps as a list of road segments, uniquely classified by ID and degree of importance. There are five degrees of importance, or functional classes, for all the road segments in Here Maps database, ranging from degree 1 for the main

highways to degree 5 for the secondary urban streets. Each road segment is also identified by the two endpoints, whose GPS coordinates are given, and by a driving direction. For each road segment, Here Maps provides also altitude information as a function of the position along the road segment.

This information was used to generate the road transportation network of the area under analysis.

### Traffic Conditions

For each road segment, Here Maps is able to provide an average traffic speed, which is the average speed that all vehicles are supposed to keep given the prevailing traffic conditions. Evidently, the average traffic speed is a function of time and can be provided by Here Maps for a desired time in the past (based on historical data), in the present (based on true traffic data provided by probe vehicles if available, or based on historical data), in the future (based on prediction models).

## 5.2.2  Vehicle Energy Consumption Model

In order to have an accurate estimation of the vehicle energy consumption, the complete vehicle powertrain must be modeled taking into account: road grade, aerodynamic and rolling friction forces, transmission ratio and efficiency, motor torque saturation, regenerative braking, electric motor efficiency, inverter efficiency and ohmic losses in the motor wirings (Fig. 5.2).

### Vehicle Longitudinal Model

The vehicle longitudinal dynamical model may be generally written as [17]:

$$m\frac{dv(t)}{dt} = F_w - F_{aero} - F_{friction} - F_{slope}$$

where $m$ is the vehicle mass, $v(t)$ the vehicle speed, $F_w$ the force at the wheels, $F_{aero}$ the aerodynamic force, $F_{friction}$ the rolling resistance force, $F_{slope}$ the gravity force.

Therefore, the vehicle model shall be written as:

$$\begin{cases} \dot{x}(t) = v(t) \\ m\dot{v}(t) = F_w - \frac{1}{2}\rho_a A_f c_d v(t)^2 - mgc_r - mg\sin(\alpha(x)) \end{cases}$$

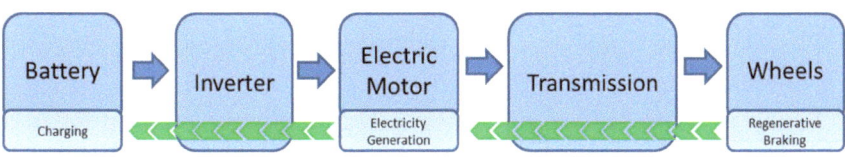

**Fig. 5.2**  Electric vehicle powertrain

where $\rho_a$ is the external air density, $A_f$ is the vehicle frontal surface, $c_d$ is the aerodynamic drag coefficient, $c_r$ is the rolling resistance coefficient, $\alpha(x)$ is the road slope as a function of the position, and $g$ is the gravity.

The sum of the aerodynamic and rolling frictions, namely the road load force, is often approximated as a second order polynomial in the speed $v$ [18]:

$$F_{res} = F_{aero} + F_{friction} = a_2 v(t)^2 + a_1 v(t) + a_0$$

where $a_0, a_1$ and $a_2$ are parameters identified for the test vehicle. Therefore, the force at the wheels can be expressed as:

$$F_w = m\dot{v}(t) + a_2 v(t)^2 + a_1 v(t) + a_0 + mg \sin(\alpha(x))$$

The torque requested to the electric motor to meet the force demand at the wheels is defined as:

$$T_m = \begin{cases} \frac{F_w r}{\rho_t \eta_t}, & \text{if} \quad F_w \geq 0 \\ \frac{F_w r \eta_t}{\rho_t}, & \text{if} \quad F_w < 0 \end{cases}$$

where $r$ is the wheel radius, $\rho_t$ and $\eta_t$ are the transmission ratio and efficiency, respectively.

An electric motor is a reversible machine, therefore it acts as a motor when $T_m$ is positive and as a generator (i.e. energy recovery) when $T_m$ is negative. The torque generated by the electric motor is saturated by a maximum torque $T_{m,\max}$ and a minimum torque $T_{m,\min}$. In particular, during braking phases, if the motor torque is less negative than the negative saturation $T_{m,\min}$, then the vehicle is slowed down only by the regenerative brake. Otherwise, the mechanical brake works together with the regenerative brake.

The power available at the electric motor shaft, in the presence of regenerative braking mechanism, shall be written as follows:

$$P_m = \begin{cases} T_{m,\max} \omega(t), & \text{if } T_m \geq T_{m,\max} \\ T_m \omega(t) & \text{if } T_{m,\min} < T_m < T_{m,\max} \\ T_{m,\min} \omega(t) & \text{if } T_m \leq T_{m,\min} \end{cases}$$

Where $\omega(t)$ is the motor rotational regime and is defined as:

$$\omega(t) = \frac{v(t)\rho_t}{r}$$

Finally, the power demand at the battery of the electric vehicle, can be expressed as:

$$P_b = \begin{cases} \frac{P_m}{\eta_b}, & \text{if} \quad P_m \geq 0 \\ P_m \eta_b, & \text{if} \quad P_m < 0 \end{cases}$$

Where $\eta_b$ is the electric drive overall efficiency (i.e. electric motor, inverter, wiring, battery).

In order to improve the model accuracy and the reliability of the energy consumption estimation, it is important to take into account also the auxiliary power requirements. In fact, the power required for driver's comfort, namely the heating or the air conditioning system, has a significant impact on the energy consumption. In particular, for an electric vehicle, comfort power requirement has a major impact on the range, and such impact increases with the trip duration and the ambient temperature difference with respect to the desired cabin temperature.

The auxiliary power term for comfort can be expressed as a function $K$ of the ambient temperature, as illustrated in Fig. 5.3.

$$P_{aux} = K(T_{amb})$$

Therefore, the battery energy consumption over the generic travel time $T$ is obtained as:

$$E_b = \int_0^T (P_b + P_{aux}) dt$$

*Macroscopic Adaptation of the Vehicle Model*

A time-variant speed or acceleration profile is not available via the map web-services, therefore the energy consumption model previously described cannot be directly adopted to compute the energy expenditure of a trip.

The model equations presented in the previous section can be rewritten by replacing the time-variant speed $v(t)$ with the average traffic speed $\bar{v}$ provided by Here Maps. All the vehicles on the road segment $i$ are supposed to travel at speed $\bar{v}_i$. As a consequence the impact of acceleration is no longer taken into account, which

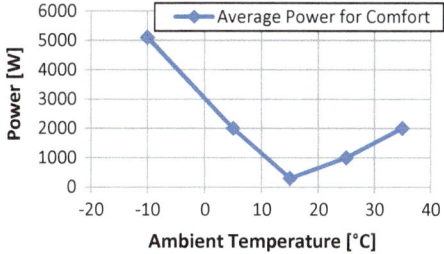

**Fig. 5.3** Auxiliary power absorption as a function of the ambient temperature

can have a non-negligible impact on energy consumption. Experimental tests conducted internally showed that this criticality arises whenever the acceleration term $m\dot{v}(t)$ is not negligible when compared to the sum of the friction and gravitational forces $F_{res} + F_{slope}$. In other words, the force at the wheels could be largely underestimated if the accelerations are neglected when the speed is low and the road grade is negative (e.g. a typical scenario of an urban traffic jam or a traffic light on a link with negative road grade). Intuitively, for an electric vehicle, which is able to recover energy through braking, this could translate into predicting energy recovery on links that are actually quite energy expensive. Therefore, the energy consumption estimation based solely on average speed is not suitable for electric vehicles.

In this work, the accuracy of the macroscopic energy consumption estimation and of the eco-routing navigation is demonstrated to improve significantly by considering accelerations induced by the different average traffic speeds and/or the known infrastructure elements (i.e. turns, signalized intersections, etc.).

**Interface accelerations**

In order to take into account the acceleration effects, and improve the model and the energy consumption estimation, the trip on each road segment is supposed to be composed of two phases: a cruising phase at the constant speed $\overline{v}_i$, and an acceleration phase to go from $\overline{v}_{i-1}$ to $\overline{v}_i$. Therefore, even though the available information imposes to neglect the temporal acceleration within the road link, the spatial acceleration that takes place at the interface between adjacent links will be considered.

The time-varying velocity $v(t)$ in every transient is linearly modeled as:

$$v(t) = \overline{v}_{i-1} + sign\left(\overline{v}_i - \overline{v}_{i-1}\right)at$$

with $\overline{v}_{i-1}$ being the constant velocity on the incoming link, $\overline{v}_i$ the constant velocity on the outgoing link, $a$ the constant acceleration to perform the jump, and $t$ the time needed to perform the speed variation.

A limitation of this modeling solution is that speed fluctuations within a link are neglected, which could be an issue in traffic. However, modeling the acceleration within a link seems unrealistic at a route planning level, because it highly depends on local traffic conditions, and as such is often spontaneous and unpredictable. Also note that this assumption mainly affects long and congested links; more specifically, it may lead to underestimation of the energy consumption on those road segments. The employed map web-service provides a densely discretized information about the road network with rather short links (i.e. average length of about 50 m). With such a configuration, it is safer to assume that the travel speed in each link is equal or close to the average traffic speed $\overline{v}_i$.

**Infrastructure-induced accelerations**

The consideration of the interface accelerations allows the energy consumption model to be more precise and realistic. However, as mentioned before, the available a priori information about the average traffic speed is not always complete nor

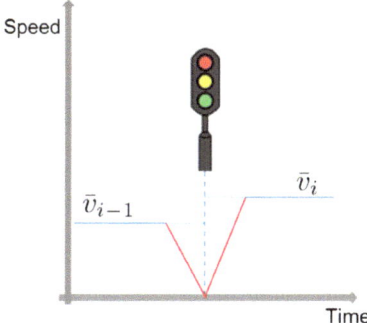

**Fig. 5.4** Infrastructure-induced accelerations considered in the vehicle energy consumption model

updated. In particular, it is very difficult and unlikely to have accurate speed information on secondary urban streets. In fact, it may happen that the provided traffic speed information on a secondary street is merely equal to a nominal value, thus not offering a speed variability across adjacent road links. This may lead to underestimations of the energy consumption, and therefore lead the eco-routing algorithm to blindly favor this type of streets. Therefore, additional information, besides the average traffic speed, should be considered in the energy consumption model for a higher robustness in case of partial available information.

Disruptions in the speed profiles and accelerations are caused not only by traffic, but also by the infrastructure. In particular, critical elements of the road infrastructure, such as traffic lights and intersections are very likely to induce stops or significant decelerations. Therefore, when the information about the position of these critical elements is available, the energy consumption term associated with acceleration should be modified accordingly.

For instance, a stop and therefore an additional speed variation could be introduced if a traffic light, or a stop sign is known to be located at a specific intersection. The speed change between two road segments connected by such intersection will be then modeled as two distinct transients: the first transient from $\bar{v}_{i-1}$ to 0, the second transient from 0 to $\bar{v}_i$ (see Fig. 5.4).

### 5.2.3 The Energy-Oriented Navigation System (Eco-Routing)

The provided map data need to be interpreted and used in order to represent the road transportation network. For the correct deployment of an energy-oriented navigation system, each road segment of the topology needs to be assigned an energy cost, which represents the energy expenditure for traveling on the segment. Such energy cost is evidently provided by the vehicle consumption model, which makes use also of topology information for a more realistic energy consumption estimation.

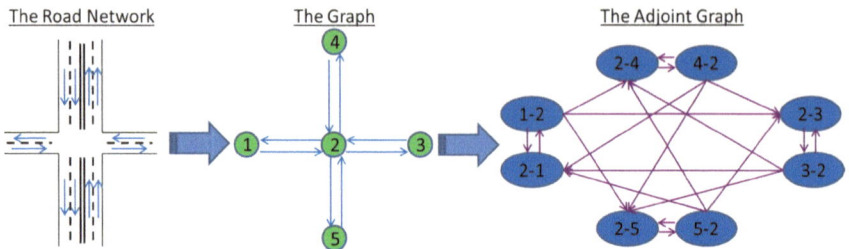

**Fig. 5.5** Construction example of an adjoint graph

### The Graph as a Modeling Tool for the Road Transportation Network

The road transportation network can be modeled as a graph, a versatile mathematical tool often used to represent any physical or chemical structure that features interconnections.

Let $G = (V, A)$ be such a graph, where $V$ is the set of vertices (or nodes) of cardinality $n$, and $A$ is the set of feasible arcs (or links) of cardinality $m$ connecting the nodes of the graph. Let us define a weighting function $w : A \rightarrow W$, which associates each link of the graph with a weight.

In conventional navigation systems and routing graphs (e.g. Google Maps), the weight associated with each arc is either the length of the arc or its travel time. In the eco-routing framework, each link of the graph is assigned a weight that represents the energy expenditure to travel on the link. The consideration of the interface accelerations between adjacent links arises a major issue in the graph modeling of the road network. In particular, every node of the graph with two or more incoming links is critical because $\bar{v}_{i-1}$ and, consequently, the transient energy consumption are not unique. Evidently, this prevents from assigning unique energy costs to the links of the graph. Hence, the graph $G$ is not adequate for the proposed energy modeling approach.

This difficulty can be resolved by introducing another mathematical tool of graph theory, often used to study the intrinsic connection properties of regular graphs: the *adjoint* graph. The adjoint graph $L(G)$ of a graph $G$ has as nodes the links of $G$, and two nodes of $L(G)$ are adjacent whenever the corresponding links of $G$ are adjacent. The routing problem will be then solved on the adjoint graph $L(G)$ instead of the original graph $G$ (Fig. 5.5).

Using the adjoint graph $L(G)$ as routing graph allows to uniquely assign the energy weights to each link of the graph, by decoupling all the possible turning movements modeled in the original graph $G$. Each link of the adjoint graph represents a path on two adjacent links, and therefore each link of the adjoint graph contains information about a link of the original graph and its upstream link.

This intrinsic property of the adjoint graph allows not only to properly consider the interface accelerations between adjacent links (i.e. $\bar{v}_{i-1} \rightarrow \bar{v}_i$), but also to model in a more realistic way the impact of the infrastructure on the energy consumption. More specifically, as previously mentioned in the vehicle model section, a stop and a modified energy term (i.e. $\bar{v}_{i-1} \to 0 \to \bar{v}_i$) can be introduced on the links of the adjoint graph that contain a critical element of the infrastructure.

### The Eco-Route Searching Algorithm

Finally, the last stage of the proposed navigation system consists in the actual search of the route that minimizes the energy consumption to travel from a selected origin to a destination in the road network. Minimization of a cost in a weighted graph can be solved by means of a shortest path algorithm.

As previously said, such an algorithm will be run on the adjoint graph. Note that the weights on the arcs can be negative, since regenerative braking is considered. Therefore, the Bellman-Ford algorithm is a necessary choice to find the global optimum [14]. Once the algorithm returns the shortest path on the adjoint graph (i.e. the most energy-efficient), the result can be easily mapped back to the original graph, in order to have the sequence of nodes composing the eco-route.

### Real-Time Traffic Conditions and Eco-Route

The road grade and the road network topology, along with available traffic lights position, as previously mentioned, are supposed not to vary over time and can be safely stored offline for faster computation. The traffic conditions, on the contrary, are extremely time-dependent. An ideal real-time query of the map data to retrieve the traffic conditions of the entire road network is not suitable for online use of the eco-routing navigation system. In fact, this would require not only the necessary time to perform all the queries for each link of the graph, but also the time for updating the weights (i.e. recalculating the energy costs) on all the links of the adjoint graph.

Another option, more suitable for online purposes, may be to store offline the global information about historical traffic conditions for different days of the week and for different times of the day. The real-time adaptation is carried out only after the driver selects the desired origin, destination and departure time (see Fig. 5.6). The desired departure time is used to load the historical traffic conditions stored offline relative to the same day of the week at the same time of the day. The N-best eco-routes are calculated on the graph weighted with the historical data using an adaptation of Yen's algorithm [19]. Their total cost is updated then according to the traffic conditions at the desired time of departure, and compared in order to determine the current best itinerary in terms of energy consumption.

Evidently, the advantage is that only a subset of links need to be updated. The drawback is that the solution becomes sub-optimal. However, it is often the case to trade an impractical global optimal solution for a faster local sub-optimal one.

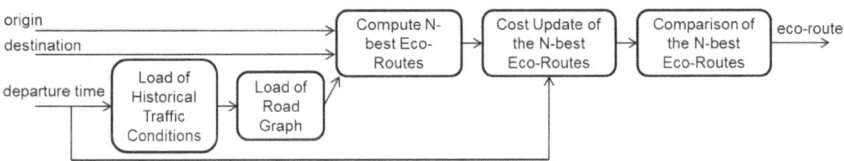

**Fig. 5.6** Block diagram of the eco-routing algorithm with consideration of current traffic conditions

## 5.3  Validation Results

The vehicle energy consumption model (described in Sect. 5.2.2) and the eco-routing strategy (described in Sect. 5.2.3) have been also tested and validated during an experimental campaign conducted in Turin at CRF on the Fiat 500e.

The objective of the experimental campaign was twofold: firstly tuning the vehicle parameters and validating the vehicle model, then comparing the energy consumption of the different routes to validate the eco-navigation system.

### Vehicle Model Validation

In the following an overall statistical analysis of the data will be reported to demonstrate the accuracy of the proposed vehicle consumption model using both the real measured driving profile $v(t)$ and the average traffic speeds $\overline{v}_i$ provided by Here Maps. The map data provided by Here Maps and used in the following analysis are relative to the week 04/04/2016–10/04/2016 at different times of the day collected every hour. The energy consumption prediction for each driving test was conducted by using the historical map data of the same day of the week as the actual driving test at the same time of the day (e.g. driving test on July 7th at 09:46 am $\rightarrow$ historical data of April 7th at 10:00 am).

Note that using the measured driving profile $v(t)$ will allow to evaluate the accuracy of the vehicle model parameters. The altitude profile used in the model for the energy consumption calculation is provided by Here Maps.

On the other hand, using the average traffic speeds $\overline{v}_i$ provided by Here Maps will allow to additionally evaluate the reliability of the modified energy consumption model, which introduces interface accelerations and infrastructure-induced accelerations, as described in Sect. 5.2.2. Evidently, the estimation based on the $\overline{v}_i$ is expected to be less accurate than the estimation based on the measured $v(t)$.

In Fig. 5.7, it is shown that the energy consumption estimation error using the measured driving speed is in average less than 5% for the entire data collection of 35 driving tests. On the other hand, the energy consumption estimation error using the average traffic speeds provided by Here Maps is in average less than 8%.

In Fig. 5.8, the travel time prediction error using the average traffic speeds $\overline{v}_i$ provided by Here Maps is reported. The average value of the travel time prediction error is 12%.

The following important conclusion can be drawn. The proposed vehicle energy consumption modeling approach takes the raw values of $\overline{v}_i$ and processes them in order to enrich the information and improve the energy consumption estimation accuracy by introducing also acceleration terms in the computation. On the other hand, the travel time estimation, which is out of the scope of the Optemus project, is carried out by simply using the raw values of $\overline{v}_i$.

The travel time prediction is less accurate than the energy consumption estimation. This means that the information about the average traffic speeds provided by Here Maps is not extremely accurate. Although the information about the average traffic speeds is imperfect, the proposed energy consumption model allows

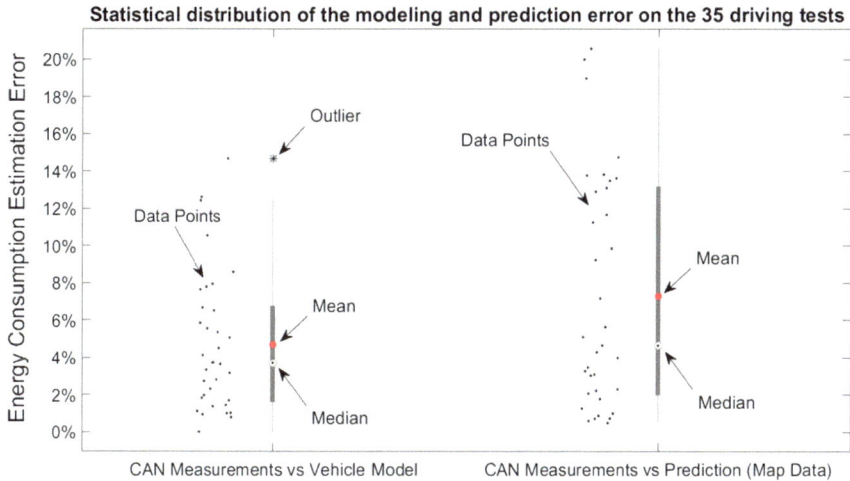

**Fig. 5.7** Distribution of the energy consumption estimation error. On the *left-hand side* the estimation error using $v(t)$. On the *right-hand side* the estimation error using the $\overline{v}_i$

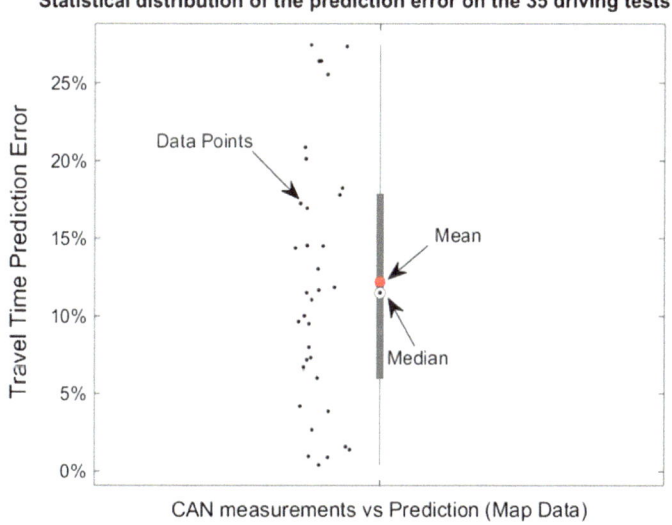

**Fig. 5.8** Distribution of the travel time estimation error using the $\overline{v}_i$

to reach promising estimation performance. Note that the map data and in particular the average traffic speeds used in the validation are historical data stored offline and relative to a sample week in the past.

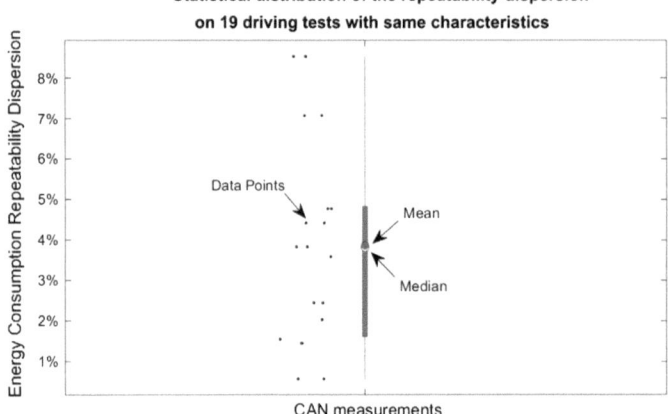

**Fig. 5.9** Distribution of the measured energy consumption repeatability dispersion

The travel time prediction error is higher than the travel time prediction error that can be observed by using the Here Maps website. This means that Here Maps computes the travel time prediction by processing the average traffic speeds through proprietary algorithms in order to improve accuracy.

Finally, a statistical analysis of the repeatability properties of the driving tests was conducted, as shown in Fig. 5.9. In particular, for each set of repeated routes with the same characteristics (driver, departure time, accuracy of the route tracking), the CAN-measured energy consumption of the single trip was compared to the average energy consumption of the set of repetitions. The repeatability characteristics of different driving tests of a same route are important for the next validation phase concerning the performance of the eco-route.

### Eco-Routing Validation

For the validation analysis of the proposed navigation system, the performance on one of the tested O/D pairs will be reported in the following.

The O/D pair that will be analyzed here is:

- Origin: 393, Corso Moncalieri, Turin, Italy
- Destination: Variante del Dojrone, Rivalta di Torino, Turin, Italy

The eco, shortest and fastest routes tested in the experimental campaign are shown in Fig. 5.10.

For validation of the proposed navigation system, let us report first the cumulative sum of the real and predicted energy consumption over the entire length of the trip. This allows not only to compare the final value of the cumulative sum, which corresponds to the total energy consumption of the trip, but also the quality of the prediction along the trip. In fact, it is important that the prediction is able to replicate as faithfully as possible the energy consumption trends observed in the

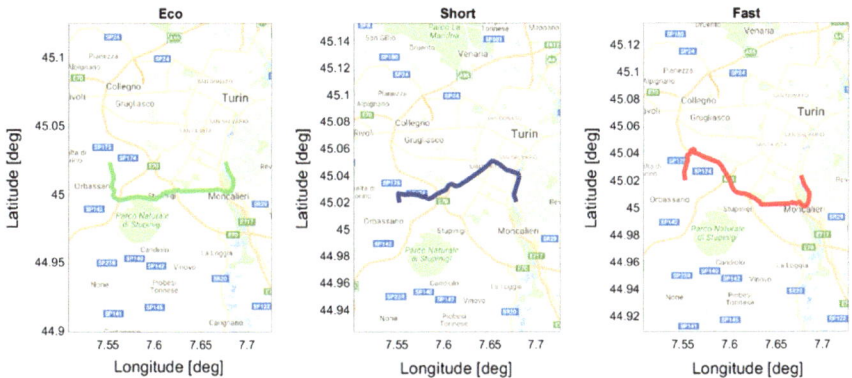

**Fig. 5.10**  The routes tested in the experimental campaign for the selected O/D pair

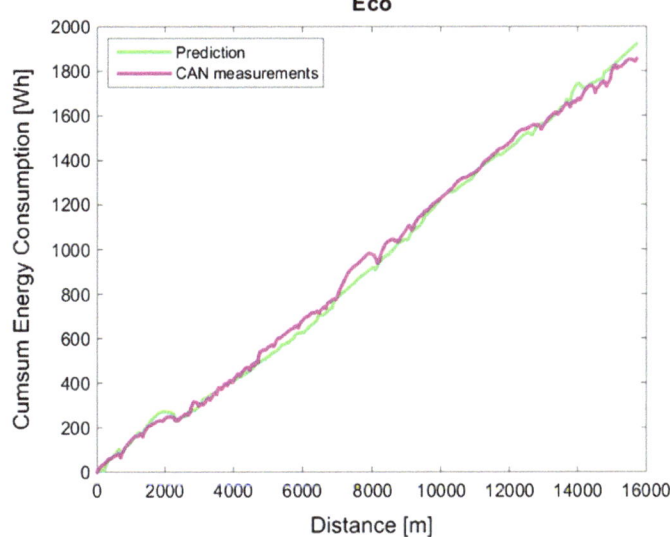

**Fig. 5.11**  Cumulative sum of the energy consumption of the eco-route in one of the repetitions. In *green* the predicted energy consumption based on the map data provided by Here Maps. In *magenta* the measured energy consumption

measured consumption. In Fig. 5.11, it is possible to compare the predicted and real energy consumption of one driving test on the eco-route.

Similarly, the energy consumption on the shortest route can be observed in Fig. 5.12.

Finally, the real and predicted energy consumptions for the fastest route are shown in Fig. 5.13. The prediction appears to make a significant error on the trip portion between 8 and 12 km. This corresponds to the stretch of highway of the

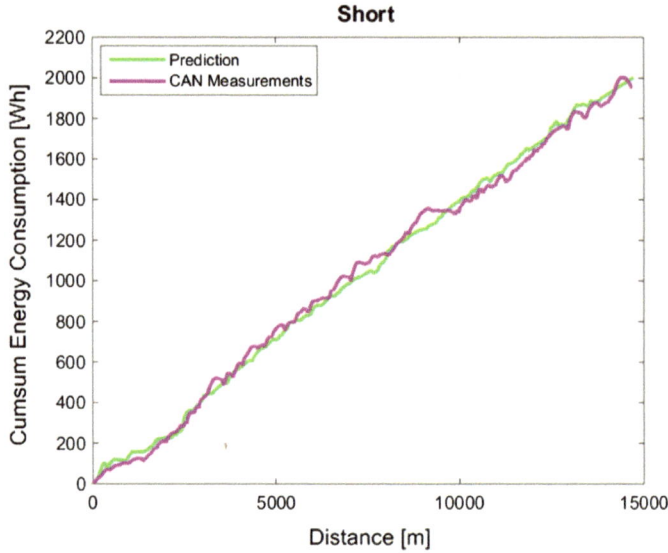

**Fig. 5.12** Cumulative sum of the energy consumption of the shortest route in one of the repetitions. In *green* the predicted energy consumption based on the map data provided by Here Maps. In *magenta* the measured energy consumption

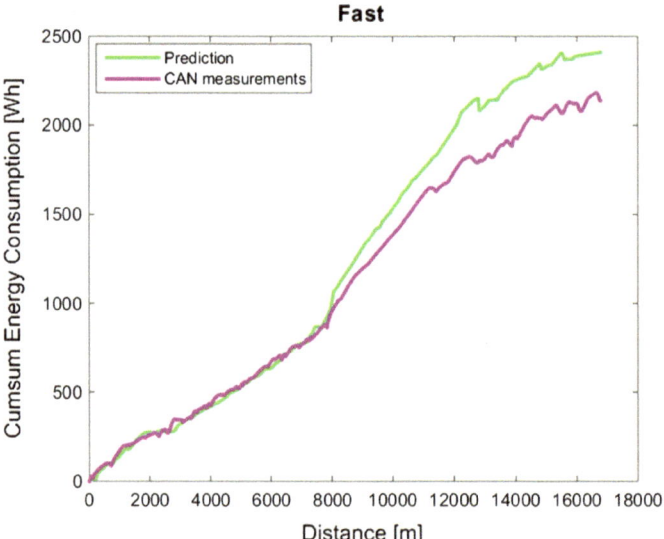

**Fig. 5.13** Cumulative sum of the energy consumption of the fastest route in one of the repetitions. In *green* the predicted energy consumption based on the map data provided by Here Maps. In *magenta* the measured energy consumption

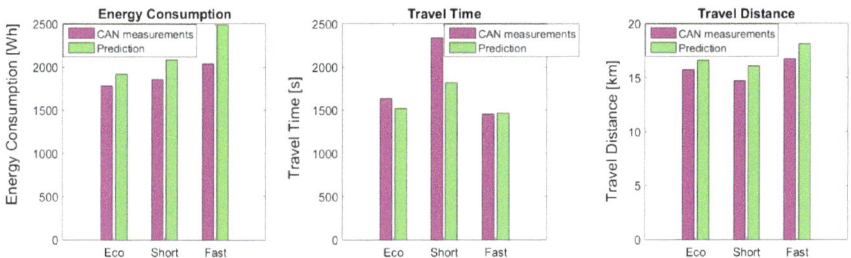

**Fig. 5.14** Measured and predicted average performance of the eco, shortest and fastest routes in terms of energy consumption, travel time and travel distance

fastest route, therefore a portion of the trip in which the travel speed is above 100 km/h. The sources of this error could be multiple and concurrent: inaccurate estimation of the road grade, inaccurate vehicle model parameters, difference between the actual driving speed and the average traffic speed estimated by Here Maps. However, the final error on the energy consumption estimation remains limited to about 13%. As demonstrated in the previous section and in Fig. 5.7, the overall prediction error is quite low.

In Fig. 5.14, the average measured and predicted performance indices of the eco, shortest and fastest routes are reported. The average is calculated over the repetitions for each type of route. It is possible to observe that the proposed navigation system predicts higher gains in terms of energy consumption of the eco route with respect to the shortest and fastest route. The measured performance gain is more moderate. Overall, it is possible to observe a performance consistency among the three routes. In other words, the eco-route is the most energy efficient route both in simulation and in reality. Similarly, the shortest route is the minimum-distance route, and the fastest route is the minimum-time route.

A focus on the average measured energy consumption is reported in Fig. 5.15. According to the CAN measurements of the driving tests, the eco-route allows to achieve 5% savings in terms of energy consumption with respect to the shortest route and 13% savings with respect to the fastest route.

## 5.4   Conclusions

In this work an innovative energy-efficient navigation system is described.

The existing eco-routing methods are inadequate to accurately estimate the energy consumption of an electric vehicle in urban environment. Furthermore, the impact of onboard accessories and systems for the driver's comfort is often neglected. This work demonstrates that the proposed vehicle model and navigation system is able not only to accurately estimate the real energy consumption of the vehicle, but also to give a reliable route suggestion to the driver to reduce the trip

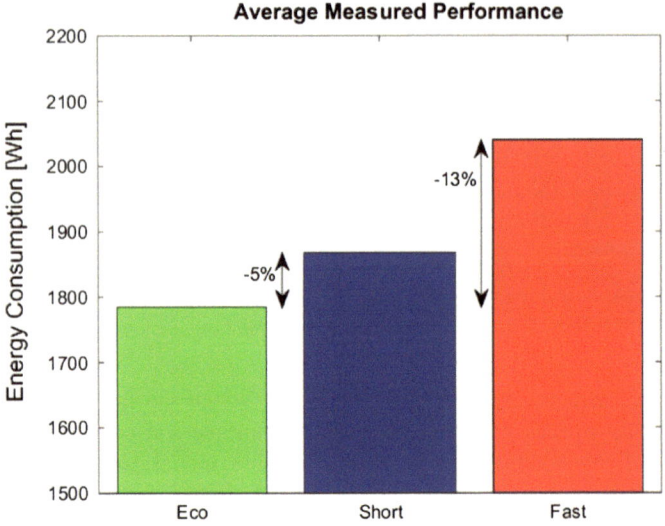

**Fig. 5.15** Average measured energy consumption of the eco, shortest and fastest route

energy consumption. This is achieved by simply using historical map data and average traffic speeds available on commercial map web-services.

An experimental campaign conducted on the demonstration vehicle (i.e. Fiat 500e) validated both the adopted vehicle energy consumption model and the proposed navigation system. Based on vehicle data measured over the CAN bus, the average energy consumption savings of the eco-route with respect to the shortest route are in the order of 5%, and the savings with respect to the fastest route are in the order of 13%.

**Acknowledgements** This project has received funding from the European Union's Horizon 2020 research and innovation program under grant agreement No. 653288—OPTEMUS.

## References

1. Sciarretta A, De Nunzio G, Ojeda LL (2015) Optimal ecodriving control: energy-efficient driving of road vehicles as an optimal control problem. IEEE Control Syst Mag 35(5):71–90
2. Ahn K, Rakha H (2008) The effects of route choice decisions on vehicle energy consumption and emissions. Transp Res Part D 13:151–167
3. Ericsson E, Larsson H, Brundell-Freij K (2006) Optimizing route choice for lowest fuel consumption potential effects of a new driver support tool. Transp Res Part C 14:369–383
4. Kubička M, Klusáček J, Sciarretta A, Cela A, Mounier H, Thibault L, Niculescu S-I (2016) Performance of current eco-routing methods. Intelligent vehicles symposium, 2016
5. Guo C, Yang B, Andersen O, Jensen CS, Torp K (2015) EcoMark 2.0: empowering eco-routing with vehicular environmental models and actual vehicle fuel consumption data. Geoinformatica 19:567–599

6.  Minett CF, Maria Salomons A, Daamen W, van Arem B, Kuijpers S (2011) Eco-routing: comparing the fuel consumption of different routes between an origin and destination using field test speed profiles and synthetic speed profiles. IEEE forum on integrated and sustainable transportation systems, pp 32–39

7.  Guo C, Yang B, Andersen O, Jensen CS, Torp K (2015) EcoSky: reducing vehicular environmental impact through eco-routing. In: 31st IEEE international conference on data engineering, pp 1412–1415

8.  Ahn K, Rakha HA (2013) Network-wide impacts of eco-routing strategies: a large-scale case study. Transp Res Part D 25:119–130

9.  Boriboonsomsin K, Barth MJ, Zhu W, Vu A (2012) Eco-routing navigation system based on multisource historical and real-time traffic information. IEEE Trans Intell Transp Syst 13 (4):1694–1704

10. Yao E, Song Y (2013) Study on eco-route planning algorithm and environmental impact assessment. J Intell Transp Syst 17(1):42–53

11. Richter M, Zinser S, Kabza H (2012) Comparison of eco and time efficient routing of ICEVs, BEVs and PHEVs in inner city traffic. IEEE vehicle power and propulsion conference, pp 1165–1169

12. Jurik T, Cela A, Hamouche R, Natowicz R, Reama A, Niculescu SI, Julien J (2014) Energy optimal real-time navigation system. IEEE Intell Transp Syst Mag 6(3):66–79

13. Nie YM, Li Q (2013) An eco-routing model considering microscopic vehicle operating conditions. Transp Res Part B 55:154–170

14. Bellman R (1958) On a routing problem. Q Appl Math 16(1):87–90

15. Abousleiman R, Rawashdeh O (2014) Energy-efficient routing for electric vehicles using metaheuristic optimization frameworks. 17th IEEE mediterranean electrotechnical conference, pp 298–304

16. HERE Maps. [Online]. Available: https://company.here.com/here/

17. Guzzella L, Sciarretta A (2013) Vehicle propulsion systems. Springer, Berlin

18. Hayes JG, Davis K (2014) Simplified electric vehicle powertrain model for range and energy consumption based on EPA coast-down parameters and test validation by argonne national lab data on the nissan leaf. IEEE transportation electrification conference, pp 1–6

19. Yen JY (1970) An algorithm for finding shortest routes from all source nodes to a given destination in general networks. Q Appl Math 27:526–530